Is Quality Just a Word We Use?

Quality management systems are essential for businesses to meet customer needs, ranging from product control to enterprise-wide process management. Effective management can elevate organizations to brand domination, while poor management can ruin an enterprise. This book equips quality experts with skills to champion business excellence and risk avoidance.

Is Quality Just a Word We Use? The Evolution from Managing Quality to Championing Organizational Excellence explores the history and flaws of quality management, offering a career opportunity for professionals that can lead to professions as expert witnesses in products liability and organizational negligence litigation. It introduces a novel quality auditing method, focusing on forensic-level investigations and case studies to illustrate the importance of prioritizing quality in business operations.

This book presents a groundbreaking model for quality professionals to drive revolutionary changes in business management, empowering them to eliminate defects and enhance their impact on business success, attracting professionals in fields such as quality assurance, quality management, risk management, and manufacturing management as well those involved in litigation.

Is Quality Just a Word We Use?
The Evolution from Managing Quality to Championing Organizational Excellence

Tom Taormina, CMC, CMQ/OE, BMSM

CRC Press
Taylor & Francis Group
Boca Raton London New York

CRC Press is an imprint of the
Taylor & Francis Group, an **informa** business

Designed cover image: Shutterstock—peterschreiber.media

First edition published 2025
by CRC Press
2385 NW Executive Center Drive, Suite 320, Boca Raton FL 33431

and by CRC Press
4 Park Square, Milton Park, Abingdon, Oxon, OX14 4RN

CRC Press is an imprint of Taylor & Francis Group, LLC

ISBN: 978-1-032-87965-9 (hbk)
ISBN: 978-1-032-87906-2 (pbk)
ISBN: 978-1-003-53562-1 (ebk)

DOI: 10.1201/9781003535621

Typeset in Times LT Std
by Apex CoVantage, LLC

This book is dedicated to my business partner and friend of 53 years, Walter "Grady" Ferguson. We shared the same values but had complementary approaches to running a business, making for a very special synergy. He was an Airborne Ranger, a pilot, a ham radio operator, and a serial entrepreneur. He was also one of the first to die from COVID-19 and his absence has been a great loss to the many who knew him.

Contents

SECTION 1   Our History to Date

SECTION 2   Organizational Pathology

SECTION 6   *Risk Avoidance*

SECTION 7   Process Excellence

SECTION 8   The Genesis of The New Quality Professions

Foreword

I was a hardcore believer in traditional quality management. The great men and women of quality, including Deming, Juran, Shewhart, Feigenbaum, Crosby, and many others, were the voices I heard when the word "quality" bounced across my thoughts.

Until recently, I was certain that the art and science of quality management had been well established. After all, we had a bag full of well-proven tools, a host of international standards, and a well-established professional society, the ASQ (the American Society of Quality), formed at the close of World War II, which has an immense global influence, with dozens of the world's countries attending their annual conference.

My confidence in quality management and our place in industry was unshakable.

But . . . deep in my bones, I knew something was missing. I remember many momentary and private thoughts asking, "Why would we assume that anything less than perfection was 'acceptable'"? I recall hearing the term "acceptable quality levels," abbreviated AQL, and thinking, "You're *expecting* a certain amount of failure."

When I was a teenager, my father, Donald Dewar, a primary founder of the quality circle movement in the West during the late 1970s, told me the story of an American company that contracted a Japanese supplier for 10,000 units, and in an effort to test their manufacturing prowess informed them that they have very strict quality standards. How strict? They would accept nothing less than 99.7% quality. That was their AQL. Period.

When the shipment was delivered, it included a single box with three units in separate packaging and a note boldly stating "Defective pieces, as per your requirement. Not for use."

That story has haunted me for decades. Then I met Tom Taormina, who spoke and wrote of the possibility of a defect, error, or mistake *never* reaching a customer. Of *avoiding* risk instead of just managing it. Of integrating quality into the enterprise's processes so they are no longer thought of as a quality-related process. Of turning it into a core business process instead of making it just a "department." These are the kind of epiphanies that create gurus, of which Tom is a member of that exclusive club.

Tom possesses the depth, credentials, hands-on experience, and theoretical framework for writing this book. In our publication alone, he's been a regular (and very popular) columnist for Quality Digest since 1996. For going on three years now, we have been working with his team to refine a professional certification program for a new generation of quality leaders. I can speak from personal experience: in going through the curriculum design process with Tom, I have had to deeply question some of my own closely held beliefs about the role of quality management in an organization. To say the least, it has been an education!

This book is a guide to the future men and women of business, education, and government. It is fundamentally a book about leadership. About doing more, better— with less. About excellence.

Among members of the quality profession, the word "quality" is normally confined to one of the typical definitions of the word. However, there is something more to it, something . . . noble. It is about doing good. It is about saving lives. It is about enriching the lives of the workforce, who typically spend their days frustrated with management's often capricious attitude toward their time, their work, and their output. This book points the way.

Jeff Dewar, CEO
Quality Digest

Preface

The definition of "quality" in products and services has been evolving since the beginning of the Industrial Revolution, and so have the methodologies of ensuring a desired and predictable outcome to any process. Over decades, we have created professions around these methods that have advanced from culling acceptable goods from defective ones (Quality Control) to attempting to avoid defects (Quality Assurance).

We have endeavored to implement the teachings of W. Edwards Deming with his *14 POINTS FOR TOTAL QUALITY MANAGEMENT* and Phillip Crosby's philosophy that *QUALITY IS FREE*. We have strived to achieve acceptable quality levels through Walter Shewhart's *PLAN, DO, CHECK, ACT CYCLE*. Those tools have become the foundation of Quality Management, which is the currently accepted methodology for minimizing defects. With each iteration, the scope of the methodologies grew and became more intricate. In our eagerness to define what quality is, we overlooked the fact that the only reasonable definition is *fitness for intended* use, nothing more, nothing less.

Or did the professions of quality take on lives of their own and continually evolve our tools and definitions to create their own culture and sub-groups within an organization? Did business owners ask us to develop more robust quality management tools, or did we sell them on how we could improve processes and enhance quality if they would authorize us to implement the latest fad?

The answer is that we have over-engineered our profession and created behemoths of size and cost with the common goal of minimizing opportunities for defects to occur. Enlightened business leaders no longer see the cost/benefit they were expecting from whatever quality conformance system that was implemented in their companies. For one example, ISO 9001 certifications peaked in the mid-2000's and are now less than 30% of that number in the United States. That decline is the basic economics of defining a return on investment, which classic quality methods have almost never achieved.

At the same time, enlightened quality professionals have seen the handwriting on the wall and are facing a bleak future as professionals and auditors. There is a disturbing article in Exemplar Global's *The Auditor* entitled "The Decline in ISO 9001 Certification: Does Quality Matter Anymore?" by Julius DeSilva. It is a reality check on the future of ISO 9001. Under international pressure, ISO announced in July 2023 that they had begun the lengthy process of revising the 2015 standard, which should be published by 2030. Technology and industry demands will change a great deal before another revision is published. Do we just hang back and wait? Quality Digest, Kaizen Online, and I have formed Q-Pro. It has published *BMS 9001:2023—A Business Management System Guidance Document* to fill the vacuum, translate ISO-Babble into business terminology, and expand the Standard to an enterprise-wide management system.

Six Sigma is also on the decline. A powerful analysis of "What ever happened to Six Sigma" was in the September 3, 2019, copy of *Quartz* magazine. The reason is basic economics, cost versus value. If you are among the thousands of "belts," savvy CEOs and CFOs have been studying the return on investment, and they are not supporting these programs any longer. The only way Six Sigma is still hanging on is the incorporation of Lean and Agile, further obfuscating the issues with the culture.

What is missing from our current quality management models will cause us to become even more irrelevant if we continue with our focus on quality tools rather than on overall business success. We will also fail if I ignore the mandate of "fitness for intended use" as the only meaningful definition of quality.

Our new role is to lead organizational change from the macro level instead of attempting to sell the relevance of quality management. We have the tools to design business models free of foreseeable risk and are committed to business excellence at all levels.

We are at a unique moment in history where quality professions can lead to evolutionary change in business management. This book presents the model for us to become pioneers in removing defects instead of minimizing them and growing our contribution to business success instead of becoming an anachronism.

About the Author

With a perspective cultivated from having worked with more than 700 companies, **Tom Taormina** brings a unique talent for precision problem diagnosis, strategic thinking, and effective written and verbal communication skills to his work as a quality consultant, trainer, and liability avoidance expert.

Tom was a member of the team at Mission Control in Houston during the Apollo 13 disaster and recovery. His experiences during his 14 years at NASA formed the foundation for his ability to analyze problems and diagnose solutions rapidly.

Tom was part of the early evolution of Quality Control Engineering at NASA. Beyond writing standards for process excellence, he also pioneered early break-throughs in supply chain management. This skill set makes him an effective business consultant and insightful expert witness in products liability and organizational negligence.

As a business consultant, Tom conducts strategic business assessments, root cause analyses, process planning, quality management, and risk assessments for companies, from start-ups to global brand leaders.

As a consulting and testifying expert in products liability and organizational negligence, he conducts formal assessments of the defendant companies to determine if they exhibited an appropriate or negligent standard of care in introducing their products or services to the stream of commerce. Tom's proprietary methodology has been trademarked Forensic Business Pathology™ and has been an accepted credential by trial attorneys and judges.

Tom has written 12 other books on the beneficial use of ISO 9001 and on risk avoidance.

Section 1

Our History to Date

1 The Way It Was

Before the mass population migration from rural environs to cities, craftsmen and tradespeople were the purveyors of most goods for the United States. Each produced products and services that were either fit for use, minimally acceptable, or of enduring value. At best, they were successful by providing commodities that consumers needed and were willing to pay or trade for. In other words, consumers determined acceptable quality by what they bought and from whom they bought it.

As the population exploded, there became a need for mass production of goods, See Figure 1.1 and factories rapidly replaced individual craftsmen. The so-called *factory system* evolved pragmatically as the new labor barons attempted to create products with minimum defects at the lowest price. With multiple people performing similar and sequential processes, craftsmanship transitioned to workers having to be monitored by quality control inspectors. Primitive American quality practices evolved in the 1800s as they were shaped by changes in evolving production styles. This methodology required the repurposing of most craftsmen, along with adding overhead inspection forces that caused problems in recruiting and retaining competent personnel.

Perhaps Frederick W. Taylor offered the first evolutionary quality paradigm. His thesis was to increase productivity without increasing the number of skilled craftsmen. He achieved this by assigning factory planning to industrial and manufacturing engineers and by using craftsmen and supervisors as inspectors and managers who executed the engineers' plans. Taylor's change in basic assumptions led to limited rises in productivity, but the new emphasis on throughput had a negative effect on quality. To remedy the quality decline, factory managers created inspection departments to keep defective products from reaching customers. The result can be assigned the distinction of the first failed quality management methodology.

World War II created a demand for military supplies that were defect-free. The practice of 100% inspections was too cumbersome and time-consuming to meet the quantities needed for the war effort. To increase output, the military invoked structured sampling practices. This effort to improve the yield rate was not a change in thinking but an experiment that had many downfalls. MIL-STD-105 became the specification for sampling by attributes. There is no published data that I can find about how statistically acceptable sampling plans were for avoiding catastrophic failures. However, MIL-STD 105 is now at revision E and contains pages and pages of sampling plans that are still in use today.

The next paradigm-shifter of the quality world was Walter Shewhart's methodology for statistical quality control (SQC) mathematics. He successfully brought together the disciplines of statistics, engineering, and economics and became known as the father of modern quality control. Shewhart determined this data could be analyzed using statistical techniques to see whether a process is stable and in control or not. An example of this is shown in Figure 1.2. This technique is still being taught in colleges and universities today and is in use in many companies.

DOI: 10.1201/9781003535621-2

FIGURE 1.1   The Early Days of Mass Production

The photo depicts a 1900s-era assembly line, often called a sweat shop.

W. Edwards Deming became a proponent of Shewhart's SQC methods and later became a leader of the quality movement in both Japan and the United States. His body of work further reinforced Shewhart's paradigm shift. Deming, however, became disillusioned with the SQC methods when WWII ended and government mandates came to an end. Demings most profound success was influencing business leaders with his methodologies, not creating breakthroughs in SQC. Deming and Shewhart both have well-deserved prizes named for excellence named after them, however.

Another acknowledged leader in quality is Joseph Juran. In 1979, he founded The Juran Institute, whose mission is to "Create a global community of practice to empower organizations and people to push beyond their limits." Again, this was not a dramatic paradigm shift but a slow and methodical evolution of best practices in quality management. In fact, these three celebrated leaders were more effective at normalizing quality management practices than they were revered for break-through changes. Their accumulated work became what we now call Total Quality Management (TQM). That is a management system for a customer-focused organiza-tion that involves all employees in continual improvement. I will characterize TQM as the sequential quality paradigm shift that is still employed in many industries today.

The last significant quality paradigm shift was the release of ISO 9000 in 1987. It did away with strict military and British standards and replaced them with a more benign methodology of process and standards conformance. I use the word benign because the practice of internal and third-party auditing that evolved with the Standard was non-confrontational and an "open book" methodology for determining process conformance.

The move away from the British and U.S. military quality standards was one of the greatest paradigm shifts in the evolution of quality management. Early on, I saw the potential for ISO 9001 to become the underlying foundation for almost any

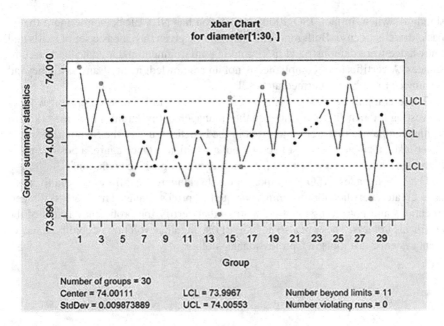

FIGURE 1.2   A Typical Shewhart Control Chart

Walter Shewhart pioneered the concept of creating bar charts, which are typically used to plot the outcome of a process. Each dot represents one component of a production run. The vertical axis shows the nominal dimension in the center and then shows the upper control limit and the lower control limit. Components outside the UCL and LCL are defective and must be removed.

quality system in any industry. It provided structure while encouraging continuous improvement. As I set off into the world of consulting in 1991, I was certain that organizations would embrace the ISO 9000 standards as a new beginning for pro-active quality management. I also hoped that I would have more work than I could manage.

At about the same time, large petrochemical companies, among other segments, saw ISO 9000 certification as a viable tool to qualify suppliers. It was not long before every oilfield supplier was clamoring to become certified.

At that moment in time, there were a handful of registrars working with notified bodies and credentialed consultants. Within the ISO boom in the oilfield, there was suddenly a flood of registrars that were not affiliated with notified bodies. There were those who could barely spell ISO 9000 hanging their shingles out as expert consultants. Most only had experience at the company they worked for before offering services to a wide range of companies in diverse industries.

These so-called consultants created a model of preparing companies to become certified in a few months without fully embracing the proactive model of continual improvement. Suddenly, factories were hanging certificates on the wall in company lobbies and flying banners from the rooftops proclaiming certification to a "higher standard" when, in fact, ISO 9000 certification has NEVER been any more than a basic driver's license. Both represent that, on one given day, a basic set of skills and knowledge were demonstrated in a sampling environment under controlled circumstances. A certificate of compliance is not an acknowledgment of any skills beyond the novice to, at best, intermediate level.

These companies spent massive overhead dollars to become compliant without harvesting the benefits of a viable quality management program. I had to work hard to find companies that were interested in the beneficial use of the Standard. I would never take an engagement just to crank out a certificate. Once again, a positive paradigm shift deteriorated over time into another not-so-positive evolution.

Over the decades, ASQ and other reputable standards and training organizations have created certification programs for quality professionals. These are skills certifications and not professional credentials that certify the proficiency levels of the certificate holders. They are also not part of the ongoing evolution of quality management systems, just tools in a professional's toolbox.

2 The Way It Is

Present day, we have the methodologies to ensure a desired and predictable outcome for any process. We have created professions around it that have advanced from QC to QA to TQM to our current forms of quality management. With each iteration, the scope of the methodologies grew and became more challenging.

As businesses, processes, products, and services become more complicated, the compelling need to minimize defects and errors continues to grow. The quality professions continue to evolve more and more sophisticated quality management systems to fill the perceived needs of their businesses.

The Six Sigma community is so self-invested that they are not looking at what is happening to companies like GE, which led the phenomena and are now in decline. You can fade into history like quality circles and Jack Welch, or you can expand your horizons and potential by paradigm shifting from the traditional tenets of quality management.

While there are clear signs that the current quality management models are no longer serving many businesses, ASQ membership has been in decline, and many quality professionals are in denial, along comes a pandemic that drastically disrupts the global economy. Figure 2.1 shows that, along with the decline in ASQ membership, the number of ISO 9001 certifications has been on the decline since the start of the COVID pandemic.

Business leaders had been scrambling to devise methods to keep their companies open and operational. Some can keep their deliverables flowing through their peoples' home offices. Others employed social distancing and other safety guidelines to keep operations going at a scaled-down level.

The pandemic has dramatically expedited the decline in the ongoing shrinkage of the need for quality professionals. If you are an internal quality person, you are looking over your shoulder, wondering when the layoff notice is coming or planning for early retirement.

The immediate lesson from the perfect storm of 2020 is that we are, as a nation, unaware of the potential risks that are looming within and beyond our control. Also, within the quality world, we are just now including risk in our lexicon, but we really do not have a consensus on what risk is and what our role is in removing risk. We must now consider:

- Lessons learned.
- Awareness of who we are.
- Conformance and continual improvement.
- We have our tool set.
- We are not on the same page as Management.
- We are often an overhead expense.

DOI: 10.1201/9781003535621-3

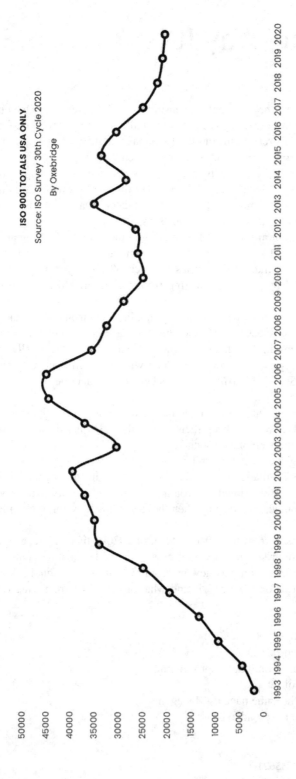

FIGURE 2.1   The Decline of ISO 9001 Certificates

This chart plots the number of ISO 9001 Certificates in force in the United States from 1993 to 2020. It shows that the number of certificates still valid has declined precipitously from its peak in 2006.

3 The Way It Must Be

3.1 ADOPT A NEW LEXICON

The following are terms I will use throughout the book that should be incorporated into your quality lexicon:

Business Management System—Enterprise-wide system of managing all business processes under one coherent infrastructure.

Business Pathology—The study and diagnosis of business systems that go beyond traditional root cause analysis and discover any foreseeable risk and potential liability issues.

Business Process Management—Every activity is a process. Every process needs a defined structure. Every process must be integrated into the Business Management System.

Business Process Mastery—Demonstrated ability to master processes and their relationship with all other processes at a high level of mastery.

Certification—Receiving written acknowledgment from an authoritative body that an individual has completed the required coursework and successfully passed all exams within the scope of any given course.

Critical Defect—As defined by your organization and culture, a defect that can never reach a customer.

Forensic Investigation©—An enhanced version of proactive management system auditing. It includes evaluations of business process effectiveness and utilizes the tenets of foreseeable risk to uncover opportunities for risk avoidance. It also can be used as a powerful tool for competence evaluation.

Forensic Business Pathology (FBP)® The highest level of business process mastery. The term means: Forensic—The application of scientific investigative methods and techniques. Business Pathology—Identifying and removing foreseeable risks from within an organization.

Quality Masters' Certification Program—A five-tiered program that progressively trains and certifies individuals to become the consummate Champion of Business process excellence and risk avoidance. Training is provided by Q-Pro, a unit of Quality Digest and Certified by Exemplar Global.

Organizational Negligence—A legal term that describes the degree of neglect an organization exhibited in delivering their products or services to the chain of commerce.

Pathological Organization©—A learning model for understanding that organizations are living organisms, each with its own personality and pathology.

Products Liability—A legal term defining the liability of any or all parties along the chain of manufacture of any product for damage caused by that product or service.

DOI: 10.1201/9781003535621-4

Quality as a Profit Center©—A proprietary methodology to evolve overhead quality systems into more robust systems that yield a return on investment.

Quality Management System (QMS)—A structured conformance system such as ISO 9001:2015.

Quality Professional—An individual who has five or more years of experience in quality systems and is skilled in the use of quality tools. Prerequisite for Course 1 of the Q-Pro Certification Program.

Quality Process Control—The structure of maintaining control over processes within a QMS.

Quality System—Any of the various techniques evolved over the decades to control quality outcomes.

Risk Avoidance—The breakthrough process for avoiding risk instead of managing it.

Risk Management—Any of many schemes that have been designed to manage risk at all levels. Traditional risk management is typically prescriptive and reactive instead of strategic and proactive.

Six Sigma—A set of management techniques intended to improve business processes by reducing *the probability* that an error or defect will occur. It assumes a certain number of defects will always occur.

Strategic Thinking—An intentional and rational thought process that focuses on the analysis of critical factors and variables that will influence long-term success.

Situational Awareness—Being aware of the environment, comprehending the situation at hand, and projecting future status with the goal of removing human error Tort-Civil litigation.

3.2   CONSTANTLY CHANGING BUSINESS CLIMATE

"The Microsoft 2022 Work Trend Index," a study of more than 31,000 people in thirty-one countries, discusses this extraordinary workforce disruption. The report found that 43% of the workforce is considering leaving their jobs in the coming year. One of the biggest reasons why people are leaving? It's not pay, the study claims. It's an unsustainable workload.

The most compelling data was the increase in time spent "collaborating." For example, between February 2020 and February 2022:

Weekly Teams meetings increased by a whopping 252%!
Six billion more emails were sent (2021 Trends Report).
We were chatting 32% more frequently.
And the average after-hours work increased by 28%.

During the pandemic, and especially during times of quarantine, our priority structures were ruthlessly simplified. Some days, our only focus was to stay alive and keep our loved ones safe. This rise in uncertainty caused an increase in mental illness. Last year, 4 in 10 adults in the U.S. reported symptoms of anxiety or depressive disorder, up from 1 in 10 adults who reported these symptoms from January to June 2019.[1]

"Job Burnout" vs. "Quiet Quitters": Job burnout and quiet quitting are two different phenomena that can occur in the workplace, and both increased during the pandemic. Burnout is a syndrome that results from chronic workplace stress that has not been successfully mitigated. It is often caused by workers having too much on their plate and is characterized by low energy or exhaustion, feelings of cynicism about one's job, and lower productivity.

On the other end of the spectrum are workers who feel disenfranchised, don't have enough meaningful work, and often hold their team back from success. Quiet quitting is a voluntary and intentional response to unmanageable workplace stress. It is often described as not actively going above and beyond at work or doing only the bare minimum to remain employed. While burnout is a consistent and prolonged state of being, quiet quitting is a coping mechanism that can be used to deal with burnout. It is important for employers to recognize the signs of burnout and quiet quitting early on.

3.3 TECHNOLOGY

Artificial Intelligence (AI) is a field of computer science that focuses on creating intelligent machines that can perform tasks that typically require human intelligence, such as visual perception, speech recognition, decision-making, and language translation. AI is a broad field that includes several subfields, such as machine learning, natural language processing, and robotics.

Spatial computing is a technology that enables computers to understand and interact with the physical world in a more natural way. It combines virtual reality, augmented reality, and mixed reality to create immersive experiences that can be used for training, simulation, and entertainment.

Cloud engineering is the practice of designing, building, and maintaining cloud-based systems and applications. It involves working with cloud platforms, such as Amazon Web Services (AWS), Microsoft Azure, and Google Cloud Platform, to create scalable, reliable, and secure cloud solutions. Together, these technologies are transforming the way we live, work, and interact with the world around us.

Many people who don't understand these new technologies are fearful of them, which, in turn, can lead to job burnout by trying to do too much to save their job or quiet quitting in the belief that their pink slips are just around the corner. The Quality Professional needs to get ahead of the curve and embrace these as new tools in their quivers.

NOTE

1 From The Harvard Business Review—The Pandemic Changed Us. Now Companies Have to Change Too.

Section 2

Organizational Pathology

4 Organizational Evolution

Let's discuss the stages of maturity an organization goes through. Most companies are launched by one or more entrepreneurs who have potentially breakthrough ideas for a new product or service. When they decide that their plan is viable, they develop prototypes or concept documents. Figure 4.1 is a graphic pyramid describing the evolutionary stages that a company can go through as its company culture matures.

4.1 EMBRYONIC

I've labeled this the "embryonic stage" of organizational development. That is, we have something in a rudimentary stage that shows potential for development. It has no final form or structure, but it will grow based on its genome and how well it's nurtured.

4.2 INFANCY

During the infancy stage of most companies, the founders do everything: product development, fundraising, finding markets, locating facilities, and doing the work.

If they're fortunate, someone's uncle or aunt does the bookkeeping. The founders' time is completely consumed with starting cash flow, and they're totally exposed to risk and liability.

There is almost never a strategic plan for the evolution and infrastructure of the company, and seldom budgets or timelines. Here is an illuminating example: When I interviewed entrepreneurs about the business plans they created during a local college program, each was obviously only a ruse to get funding. When I inquired further, the principals said they had no intention of following the pitch plan after they got funding!

4.3 CHILDHOOD

During the childhood stage, the leaders have evolved in what direction the organization is headed and have created a basic pragmatically formed infrastructure. Unfortunately, not everyone shares the same embryonic DNA. There is typically no formal vision and mission, often poorly documented processes, and no specific roles and responsibilities. They do know they need help. But, the immediate priorities are finding people to do all the needed activities, such as design and development, purchasing, production, sales, and so on. In their immaturity, they seldom see needed structure between departments and processes.

DOI: 10.1201/9781003535621-6

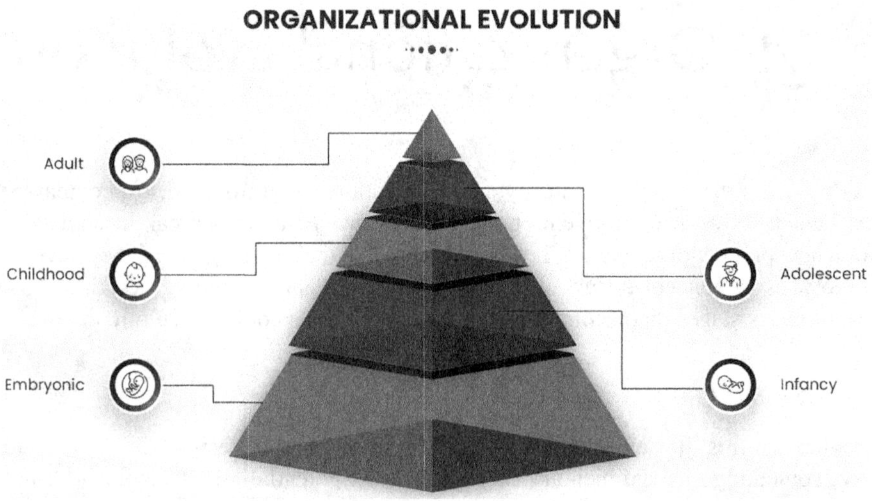

ORGANIZATIONAL EVOLUTION

··●●●··

FIGURE 4.1 Organizational Maturity Model

This pyramid depicts the growth of an organization from embryonic to adult. It represents the business pathology model in this book.

4.4 ADOLESCENCE

During the adolescence stage, the organization's culture and identity have formed, but there is no structured set of processes. The organization may also harbor unknown elemental flaws, either dormant or introduced from outside, that may infect future efforts. Besides the required employee manual, there may be little other structured documentation. The principals are in full-time damage control, putting out fires with no plan. I've seen organizations that continue to survive in chaos without ever growing out of adolescence.

4.5 ADULT

If a company is going to grow and thrive, sooner or later, someone must become the adult in the room. Many of our consulting assignments have started with a CEO looking for someone to bring order from chaos. The first step is a weekend retreat with all the senior managers. There, I identify possible problems and form individual treatment plans. Then, I form a strategic plan with vision, mission, and values identified.

4.6 ADULT-ONSET DISEASES

Unfortunately, some companies have conditions that will require more than first aid. Others discover their problem is inoperable, and they may not survive. Often, there are power and control issues among the staff that will prevent an enterprise-wide, long-term plan for organizational health. Since there are no universal process management models to structure a successful company, and no two companies have the same DNA, different organizations may attempt many directions over their lifetimes, and some of them may not be healthy choices.

5 The Pathological Organization

5.1 THE PATHOLOGY METAPHOR

The pathology metaphor was carefully chosen to prepare business leaders for running their businesses through a virtual CT scan. The procedure requires that everyone in the organization subject their methods and processes to clinical examination with the real possibility that there may be cancerous cells that we would rather not confront, admit to, or prefer to treat surgically. Business executives often would rather have a root canal than go through the process of self-examination of long-established business methods that they perceive as being successful, stable, and reliable. Only a precious few enlightened business leaders have seen the wisdom of implementing suit-proof companies.[1] The day-to-day issues and minor catastrophes consume all executives' available time. If only they could see the world through the eyes of an expert witness. There is compelling data that foreseeable failures are the result of poor business processes, ineffective quality management, and leaders who are unaware of potential risks. There are also too many instances where an owner/CEO has a "pet project" they are not willing to give up, which can keep the company from growing into its own and leading its industry. When this happens, another smaller, leaner company can overtake them in technology and market share.

Business leaders who have been bludgeoned by a costly and personally disastrous lawsuit are receptive to performing introspective diagnostics on their organization. Through the pain of fresh wounds, they genuinely want to know what went wrong with the business model and how this disaster can be prevented in the future. Similarly, a person may become predisposed to start thinking proactively about health only after a first heart attack.

I have spent more than two decades implementing the tools of Forensic Business Pathology (FBP)®. Remarkably, the profound lessons learned from Apollo 13, the terrorist attacks of 9/11, the Enron collapse, and so many more are just as valid and efficacious today as they were when they happened. It is not surprising, however, that demand for these tools during litigation is greater than offering them proactively to business leaders. As plaintiff and defense attorneys discover FBP, it creates an immediate demand for adopting these winning tools that create compelling and unimpeachable testimony of process and human failure.

Stalwarts of the business world, who have outsourced production, assumed that offshore suppliers would provide compliant components and delusionally believed that their percentage of defective products was within an acceptable margin, may be on the verge of learning the meaning of "strict liability." In most jurisdictions, the company that builds a product or supplies a service is accountable for the actions of its employees and suppliers when a product (allegedly) fails and causes harm. This

DOI: 10.1201/9781003535621-7

is a Kairos moment I would not wish on anyone. Right or wrong, the defendant company and its senior managers are in for a financial and emotional rollercoaster if they ignore the tools of prevention available today. The tenets of FBP for businesses are elegantly simple:

- Develop an irrevocable company goal that no critical defect shall ever reach a customer.
- Use quality process improvement tools to map all business processes, their interrelationships, and their dependencies.
- Assess each process on its own and assign metrics for effectiveness, quality, and reliability as business Key Process Indicators (KPIs), not quality metrics.
- Dissect each process until all opportunities to identify and understand defects or mistakes.
- Objectively improve the processes until they are innately safe and reliable, and continually monitor metrics to warn of pending problems before they happen.
- Remove the concepts of defect inspection, rework, punishment, and blame from your organization.
- Replace them with individual and organizational accountability for the outcome of everyone's work.
- Continually elicit and scrutinize customer feedback for potential issues and correct them immediately and proactively as a "gift" of valuable information.

While the tenets are simple, their implementation requires uncommon courage by business leaders to assess their business for systemic disease and eradicate it with tenacity and finality. Success requires driving any signs of mediocrity from the organization just as you would remove pre-cancerous cells from the body.

Businesses need to do more than just react to symptoms. Study businesses that are currently failing because of economic woes or products liability issues and commit to driving risk from your business. Your first introduction to FBP should not be in a courtroom.

5.2 SYMPTOMS

To arrive at a correct diagnosis, the first question is typically, "What are the symptoms?" As in medical diagnostics, the patient may not exhibit any overt symptoms other than malaise, apathy, agitation, fever, or headache. Metaphorically, I can apply these same symptoms to businesses as well. Let us do our diagnostics and work from the bottom up.

Have you been losing customers in recent years? Have customers stopped buying or using your services for no apparent reason? Have you dismissed these phenomena to the economic downturn or foreign competition? Have you independently (not through your sales department) interviewed former clients to determine why they have abandoned you? Customer loyalty is hard-earned but easily lost. You may have lost your competitive advantage, and your customers may be bottom-feeding for the lowest price, but your product or service may have deteriorated in quality or performance without your cognitive knowledge. We are disposed to throw away defective products and simply shop elsewhere. We seldom report disappointment with a product to the manufacturer. We almost never hear the ticking of a time bomb until the clock runs out.

Is your customer service phone bank in continual use? Are your people handling a steady stream of calls and e-mails about product or service issues? Is your warranty repair group as busy as the manufacturing line? If you sell to distributors or retail chains, do you receive failure reports and defective merchandise, or do you just systematically credit the customer for some predictable defect level? There is an incredible wealth of information awaiting you in your customer service and warranty activities. Those who deal with distraught or irate customers develop an impenetrable shell that defends them from personal attacks. When they hear the same story repeatedly about a product failure or service issue, they will form a defensive profile that characterizes the complainers as unreasonable, ignorant, or wanting something for nothing. I've been in call centers that do not even ask what the problem is but just refund the money or send a replacement product. I have observed others who do no more than categorize the problems as customer remorse, improper use, or random failure and then never quantify the data. Each customer call, each warranty return, and each unhappy customer is a gold mine for discovering pending liability before it happens.

Does your company operate on tribal knowledge or dictatorial rule? Are each of your processes well-defined and measured, or does everything flow downhill to a toll gate during the final inspection? Are roles and responsibilities well-defined with measurable outcomes, or are you populated with a staff who tries to do whatever job they are handed? Is there someone who makes decisions on what is acceptable, or does everyone have clear guidance for acceptable workmanship and performance? If you do not employ robust process controls with specific performance criteria at each step of your operation, you are building defects in your products or services. This is not a conjecture: there is some level of liability built into a business that may not be seen until it causes harm and you face charges of negligence.

Are your employees performing below your expectations? Is "human error" a common excuse for mistakes or oversights? Are there sub-groups formed within the organization that run the day-to-day operations? Are performance reviews subjective and pay raises automatic? Is there an inconsistent consequence for sub-par performance? If you interviewed a cross-section of employees, would you get a constant definition of your standards of customer service and quality? In businesses where those in charge are managers and not leaders, the day-to-day operations are typically anarchy, and the supervisors are firefighters who are continually stomping out flames. The smoke from these fires obscures the root causes of the problems until there is a catastrophic breakdown. At that point, we complicate the symptoms by punishing the guilty. We learn expensive lessons that we do not implement.

These are but a few symptoms of dead and dying organizations that are seen in my work as a management consultant and, even more dramatically, as an expert witness. The following three sections will attempt to describe the most common and deadly diseases and their symptoms.

5.3 ENTERPRISE ENTROPY

En·tro·py (noun): The inevitable and steady deterioration of a system.

Humans, as they age, typically move slower, forget details, and have more aches and pains. Hair usually changes color, and skin that was once well-defined now sags. Businesses exhibit a different set of symptoms with the passing of time. These changes can be very subtle and imperceptible. While human aging is natural, Enterprise Entropy (EE) is an avoidable and curable fatal disease. The longer it is undiagnosed and untreated, however, the less the chances for recovery.

Regardless of the size or origins of an organization, the dormant EE virus is present in all businesses from the day the doors open. As with most cancerous maladies, the destructive cells are imperceptible at first and may grow out of control and become terminal before they are detected.

I did not recognize EE as a distinct malignancy until I began working as an expert witness in products liability. I have witnessed the entropy of many large companies that were once healthy giants, now diseased and dispensing toxic goods and services. I have been compelled to become a skilled doctor of entropy and focus my energies on diagnosis and surgery rather than prevention. There is a spreading epidemic of this cancer, especially acute in the manufacturing and service sectors. In the examples that follow, I will construct an EE example of a product and then diagnose the symptoms (similarly applicable to services). This uses, in part, a white paper authored several years ago because this topic is so essential to understanding the quantification of standard of care and foreseeable risk.

EE can be more fully appreciated as a specific enterprise malignancy after one has worked as an expert witness in products liability litigation. Organizations that suddenly discover their malady from products liability litigation are suffering the consequences of excruciating pain. What follows is an EE pathology example, which I will examine to determine what the symptoms reveal.

An entrepreneur consumes himself with a new idea to build "a better mousetrap." He surrounds himself with like-minded friends and family who usually support the concept or the individual. They collectively make heroic efforts to successfully bring the concept to fruition and release the product for initial production and sale. Unbeknownst to all who are involved, a parasitic organism has already attached itself to the business.

In the early days of organizational development, the inner circle of entrepreneurial individuals who started the business shared a vision of what they were to build. They typically had a common concept of what the product would do, how it would perform, and what "acceptable quality" looked like. They shared some expectations of how the world would view and accept its product.

As the days and weeks of product development passed, they unconsciously decided to make changes and select one approach over another to move the process forward. Each person in the development team lived the evolutionary process and has unconsciously filed invaluable lessons about mistakes made and details of the logic used to make tactical business decisions. This was effective, but the potential to learn from reflecting on the decision-making process was lost in the problems of the next day. Gradually, the original concept morphed into something that they consensually concluded was marketable, profitable, and reliable. That production-ready final design may not have borne much resemblance to the original conceptual vision.

With good fortune, the original product evolved to new-and-improved models, spin-offs, creative applications, complementary products and services, and a viable business. Both the founders and start-up team consume themselves in the business of running a business, or they have started hiring new employees to run the day-to-day operations of the new enterprise. From the outset, their shared vision and values are diluted by bringing outside people aboard who were not present at birth and delivery. There are too few entrepreneurs who defined and codified their ideas and value systems and systematically trained each new person in the fundamentals of the intended culture of the organization. At this point, the EE virus has erupted from dormant to active.

Entrepreneurs, consciously or unconsciously, imbed their personal values in their new enterprise and the development of their products and services. This may amount to informal mandates such as: all products are tested 100%; notifying the customer in writing of any potential schedule change; freight charges being included in the quoted price; alcohol consumption is prohibited during business hours; falsifying documents is grounds for immediate dismissal; the president approves all shipments.

While the values underlying enterprise management activities are the lifeblood of the founding fathers, new employees seldom internalize them. There is rarely "time" to perform this awareness training. For example, new workers may come from companies that charge customers for shipping, and a beer at lunch may be okay. The first symptom of EE is then the lack of a shared vision and values.

From this benign first indication, the disintegrating spiral of entropy gains strength from an ongoing culture of overlooking the fundamental rules that made the business or product viable. As each new day unfolds, processes are continually changing due to improved methods, availability of raw parts, variations in demand, and a host of other demands and alterations. Each change happens with the best of intentions, but as companies grow and products evolve, the complex interaction of processes often goes unnoticed until a major symptom of entropy appears.

From products liability disasters, let's look at an example of the EE process in action. The "on-off switch" that is included on most electrical appliances may appear to be a common, simple, and reliable device. There are certain inherent safety issues in switches because there is predictable electrical arcing when switch positions are changed. The cause of the explosion and fire on Apollo 1 was attributed to this phenomenon. Years ago, a phenolic that was a very thick and robust plastic formed the housing of the switches. One of the key features of phenolic was that it did not spontaneously burn, so any type of electrical spark inside the switch had no chance of causing the housing to catch on fire. Older switch mechanisms were made of many component parts that were rugged and reliable. In many companies, successive engineering groups made the switches less complex and less expensive to build. They dutifully replaced the traditional inner workings with less expensive metals, fewer moving parts, and simpler methods of attaching wires. Contacts became less rigid and robust, so internal arcing became more prevalent. Wires contacted some minimally resilient spring mechanisms instead of hard connections like soldering. At the same time, engineers replaced the phenolic housings with more inexpensive plastics.

Most of these changes took place over many years. Each transpired with the acknowledgment of the company management and was acknowledged as a "breakthrough" idea. On their own, many of the changes were reasonable. Now, however, switch manufacturing has gone offshore, and no longer controls the consistency and quality of the integral parts and the manufacturing processes. The result has been that a once "bullet-proof" switch assembly now frequently has internal workings that are not as secure, causing electrical arcing onto a plastic material that not only burns but drips flaming hot plastic onto other flammable parts. If the original switch designers could see the "new and improved" switches, they would be shocked because of so many induced variables and the lack of testing against the original design criteria, resulting in a dangerous product unconsciously delivered to the public. For those who analyze processes and their changes, the metaphor that comes to mind is that if you place a frog in a pot of boiling water, it will jump out and save itself. If you put the frog in cold water and gradually bring it to a boil, the frog will boil to death.

The previous example of an electrical switch has been the root cause of life-taking events in three lawsuits I have worked on in recent years. These $0.25 components have collectively cost the three manufacturers in lawsuits nearly $50 million in compensatory and punitive damages. Do you have this money budgeted in your company's financial statement? Are you the next victim of Enterprise Entropy?

In working with plaintiff and defense attorneys, I have evolved the process of Forensic Business Pathology®. While other experts are performing forensic analyses of the remains of a disaster, I examine the business and process management systems and history of the defendant company. Using proven management, quality auditing, and investigative techniques, I perform detailed analyses of the evolution of products and services to find spurious events that may affect quality, reliability, or safety that entered the process undetected. FBP tears down processes into their most elemental states and looks at them in multiple dimensions with unfiltered glasses. Instead of taking any parameter for granted, I use a form of Boolean logic (and/or/else) to determine if process elements have any potential for misuse or misapplication. FBP then constructs scenarios of the sequence and interaction of these processes and determines how unexpected outcomes resulted from previously stable and reliable processes. Most often, FBP uncovers irrefutable evidence of processes gone awry undetected by usual business metrics and analysis. Unfortunately, these investigations often also uncover a breakdown in accountability and stewardship by one or more elements (people, departments, philosophy, unconsciousness, ambivalence) within an organization. The serendipitous alignment of one or more of these undetected problems, plus some unforeseen scenario conducted by a customer, is often the catalyst for catastrophic failure leading to a lawsuit.

I now successfully use the FBP techniques to help identify EE in organizations that would like to avoid products liability lawsuits. Our diagnostic tools prepare compelling and often irrefutable expert witness reports applicable to your organization to discover entropy, diagnose its symptoms, uncover the origin and locations of malignancies, perform triage, and then invoke surgery to repair the diseased processes. After the abatement of entropy, I then implement cultural changes to prevent further endemic outbreaks.

Unlike performing risk assessments within a manufacturing or service company, FBP conducts deep-dive-focused investigations on risk avoidance and how to structure processes and cultures that ensure critical defects can never reach a customer. Instead of hunting for defects, FBP eliminates defects through processes that make each person accountable for the shared goal of defect avoidance.

Business leaders need to become aware of their potential liabilities before they become the subject of a products liability suit. Traditional companies have chosen to take proactive steps to eliminate risk potential using the lessons cited above.

If successful, the original product evolves to new-and-improved models, spin-offs, unforeseen applications, complementary products and services, and a viable business matriculate from a start-up. By this time, the entropy has started because the founders and start-up team consume themselves in the business of running a business and abdicate leading employees. This almost immediately dilutes shared vision and values by bringing outside people who were not present at birth and delivery. I have yet to find an entrepreneur who defined and codified her ideas and value system and systematically trained each new person in the fundamental culture of the organization. I call these undefined attributes the "absolutely always" and the "absolute nevers." Not performing a "always" must be grounds for discipline or dismissal while performing a "never" must be grounds for dismissal.

5.4 PARADIGM PARALYSIS

Par·a·digm (noun): An example serving as a model; pattern. Closely held beliefs that guide our everyday lives and form our system of values.

Humans are more creatures of habit and environment than I would like to admit. We are the manifestation of all the experiences in our lives, and our unconscious behavior forms who we are, what we believe, and how the world perceives us. Our beliefs and paradigms manifest themselves from the greatest leaders and scientists to the most evil and destructive people in history.

I became aware of the term paradigm in the late 1980s when I was an adjunct to the American Productivity and Quality Center in Houston and The University of Houston Center for Advanced Management Practices. I was intrigued with the concept after studying the works of futurist Joel Barker.

Joel Barker was the first person to popularize the concept of paradigm shifts in the corporate world. He began his work in 1975 after spending a year on fellowship meetings and working with visionary thinkers in both North America and Europe.

He discovered that the concept of paradigms, which at that time was sequestered within scientific discussion, could explain revolutionary change in all areas of human endeavor. By 1985, he had built the case for rampant paradigm paralysis, and corporations and nations began seeking his advice. In 1980, in addition to his work on paradigms, he began to focus on a second crucial component for organizations and individuals: the importance of vision. In 1986, he released his first video, Discovering the Future: The Business of Paradigms. By 1988, it was the best-selling business video in history. It has been translated into 16 languages and continues to influence people all over the world.

He popularized the term paradigm paralysis to describe how even the greatest minds can become oblivious to alternatives to their closely held beliefs. Barker's examples include the fact that the Japanese invented the digital watch because the Swiss clockmakers were handicapped by the paradigm that watches would always have a mainspring. In the 16th century, the government imprisoned Galileo for heresy because he was teaching the Copernican theory that the Earth and planets revolved around the sun, while the leaders of the time believed that our universe orbited around the Earth. There are scores of examples of how breakthrough discoveries have come from unexpected results of scientific experiments and "outsiders" who are not invested in the beliefs of traditional industry leaders. Karl Popper[2] said, "Every good scientific theory is a prohibition; it forbids certain things to happen. The more it forbids, the better it is." The paradox of this statement is that unless we continually challenge closely held beliefs, we will continue to ignore the manifestations of new science.

This topic is extremely relevant to business pathology. We must acknowledge how the paradigms of business leaders lead to unplanned results. One industry heavily invested in its own paradigm is the legal industry. In products liability cases, business leaders suffer from paradigm paralysis about their products and businesses, while many litigators complicate these cases because of closely held beliefs in litigation strategy. I have spent decades helping industry professionals look beyond paradigms and discover breakthroughs in productivity and customer satisfaction. I have spent more than two decades helping attorneys with alternative approaches to products liability and organizational negligence tort. Stepping away from our stories and looking outside at what I believe to be true is the foundation of FBP. Humans rarely invite change, so achieving success in these arenas is a formidable challenge.

5.5 PLANNING TO FAIL

Another maxim is that if you fail to plan, you plan to fail. Business planning takes many forms and has different names in various industries. Its definition runs the gamut from how to raise money to how to bring a product to market to how to dominate a market segment. Business planning is often whatever the businesspeople believe they are doing to ensure future success.

To bring clarity to business planning, which is a poorly understood process, I use the term strategic planning. Strategic planning experts convert business objectives into a polished plan and schedule. Unfortunately, most strategic plans leave out critical elements of how an enterprise performs to its maximum potential while building in risk avoidance. Business leaders also fail to allocate the resources that effectively implement the plan being constructed. Strategic planning is too often an event instead of a way of life for daily business conduct.

Absent from most strategic plans is a process for continually communicating and enforcing the mission, vision, and values of the leadership. Many business executives acknowledge the need for a more closely-knit community within their organizations. They also put this task at the bottom of the priority list because it does not appear on the balance sheet or the P&L statement. Companies that do not communicate with everyone and enforce their value system in the organization are more likely to pay punitive and compensatory damages in a lawsuit.

Organizations without an immutable moral and ethical compass to guide every-one in the company not only achieve suboptimal business results and lower customer loyalty but have the cancer of products liability and organizational negligence metas-tasizing within them. From personal experience of performing hundreds of quality management systems audits, I have seen more quality and reliability issues among companies that did not drive quality from the top down.

Failing to plan, act, measure and improve on the vision, mission, and values of a company is planning to fail. Attorneys who employ FBP tools early in a case file have often discovered the root cause of an accident before spending exorbitant funds on expert witnesses.

5.6   IGNORING THE SYMPTOMS

Meeting with business executives as either a quality auditor or business process improvement consultants, I have seen these executives reject the facts and symptoms of entropy. Paradigm paralysis can even cause data to be dismissed as faulty or irrel-evant. The data-driven, Boolean logic, irrefutable, and emotion-free presentations eventually deteriorate into an unwanted and ignored monologue. The facts do not always speak for themselves to those who have paradigm paralysis. Worse yet, this process is for businesses that seek help and are open to constructive criticism.

To organizations with a built-in system of tolerance for mediocrity that is ineffec-tive and performs poorly with customers, data points can damage an organization's carefully crafted illusions. Not only do they choose to ignore the symptoms, but they have also devised plausible stories to defend their lack of standard of care.

The worst offenders move beyond ignoring the symptoms and are unethical or dishonest. In one case, working for the plaintiff, the data received in discovery was so redacted, disconnected, and poorly contrived that the judge allowed me to do a facility audit to verify the allegations in my expert report. One of the defendant's attorneys led the tour and continually advised company employees not to answer my questions directly. Eventually, I asked to see where they performed the flammability testing that they reported to UL. Outside the plant, they escorted us to a far corner of the property near the trash bins and pointed to a 40-foot brick chimney, which was a relic of years gone by when companies burned their trash. The attorney offered this as the company's flammability test stand.

As evidenced by the subsequent settlement, this company was hiding the fact that it had cheapened its products and circumvented long-established safety standards for the sake of market share. When confronted, they not only ignored the symptoms but justified their behavior by moving manufacturing to China.

After first beginning to help companies implement the ISO 9000 quality stan-dards, I was advised that I would receive resistance in convincing executives that they needed to consider changing processes that were not compliant with the tenets of the Standard. Our teaching model was tantamount to convincing senior manage-ment that their mothers were ugly. In other words, the company paid us to convey dis-tasteful news. Delivering audit results and proposing viable solutions are necessary to achieve ISO 9000 registration, but ignoring the symptoms is counterproductive to the positive results intended.

5.7 COMMON PATHOGENS

- **Dysfunctional Senior Management**—This refers to a situation where the senior management of an organization is unable to perform their duties effectively, leading to a decline in the organization's performance and productivity. Dysfunctional senior management can manifest in various ways, such as poor communication, lack of transparency, micromanagement, favoritism, and nepotism, among others.
- **Staff Meetings**—These are meetings that typically do not produce the desired results. They are often characterized by a lack of focus, poor communication, and a lack of clear objectives. Common examples are meetings that are too long, poorly organized, and dominated by a few individuals who prevent the agenda from being followed.
- **Manufacturing Runs the Company**—Here, the primary focus is on the manufacturing function instead of the overall company success. While their intent may be noble, it is often destructive. The goals should include the quickest time to market. How can we do it more inexpensively? While these are important goals, many organizations ignore elements like safety, quality, and employee engagement.
- **Risk Management**—This archaic concept needs to be replaced by risk avoidance. Companies should take a deep dive into their processes and define where inherent and foreseeable risk exists. Then, take proactive steps to avoid them.
- **The Supply Chain**—There are always issues with the supply chain and their management. Many were exacerbated by the pandemic. These include demand volatility, rising prices, bottlenecks, and cyber security issues.
- **Offshore Suppliers**—While having an offshore supply chain often has short-term benefits, lack of control over your suppliers and their inconsistencies and lower quality standards can lead to products liability litigation.
- **Finding the Right Employees**—The workforce is aging, and the next generations of workers are not replacing their talent and work ethic. Again, this pandemic-induced change, coupled with the decline in the desire to have meaningful careers, is a major conundrum for many companies.

NOTES

1 Referring to lawsuits.
2 Author of *The Logic of Scientific Discovery.* https://staff.washington.edu/lynnhank/Popper-1.pdf

6 From Defects to Tort

An individual can be devastated after being handed a complaint by a process server or officer of the court. The obligatory script includes a phrase such as, "You are being sued. Sign here." The executive named in the suit is stunned and gasping for air. Receiving a legal form containing archaic typeface with legal jargon buried under time stamps and case numbers can be as mortifying as when a person is publicly humiliated among their peers.

This description is intended to be graphic because that person's life is never the same after that moment. The trauma may be as devastating as losing a loved one. The ensuing days, months, and even years will be uncharted waters, subject to devastating scrutiny of issues that may not even be germane to the allegations. There will be documents containing scathing condemnations and outrageous accusations presented in indecipherable legal jargon. Large sums of money will be spent on pleadings, briefs, and filings, and there will be hours of interrogation. Even the advocates may seem like enemies, given the high cost of retainers and expenses.

If there is no settlement, the otherwise consummate business professional may be subject to the horror of a deposition followed by the torture of a trial. Since your own attorney is attempting to anticipate every question and allegation from the opposition, the preparation for a deposition can seem hostile. He or she may drill you on procedure and warn you of objections and adversarial attorneys. The scathing interchanges can leave even the most seasoned executive looking for somewhere to hide from the crossfire.

To borrow a concept from Colonel Kurtz in the movie *Apocalypse Now*, you can prepare for all contingencies in battle, but you are not ready until you have seen "the horror" and have stared it in the face. The story he told in the movie was of U.S. soldiers returning to a village where they had immunized small children only to discover that the Viet Cong had chopped off each inoculated arm. His message was that the opposition had the cunning to perform unthinkable acts of terror as a psychological device. To the uninitiated, witnessing the adversarial process of jurisprudence in the United States can be as devastating as "the horror" described by Colonel Kurtz.

This topic is included for two reasons. First, business executives must take steps to ensure the process server never visits them. Second, attorneys must be reminded that, though a lawsuit may be routine to them, their clients are ill-prepared for the unfamiliarity and not knowing the outcome. Though expert witnesses may view legal proceedings as an opportunity to hone and perfect skills, an average citizen does not look forward to mediation, deposition, or cross-examination. The advocate's job, therefore, should include painting a picture of the upcoming "horror" for clients to eliminate emotion clouding the facts.

DOI: 10.1201/9781003535621-8

7 Living the Five Stages of Grief

7.1 ON DEATH AND DYING

Regardless of whether a company's nonconformity was a complete breakdown in the system or a minor issue, those at the receiving end of a quality systems audit always exhibited, to some extent, the five stages of grief. Understanding the FBP model of assessment of business health requires a working knowledge of this phenomenon because businesses are organisms that suffer human-like symptoms in their respective lifetimes.

In her 1969 book, *On Death and Dying*, Elisabeth Kübler-Ross offered a model for how humans behave in the face of tragedy. Psychologists use the Kübler-Ross model and more commonly describe it as the five stages of grief. They are denial, anger, bargaining, depression, and acceptance. I will put these in an FBP context.

Denial is the typical reaction when an executive confronts a catastrophic business issue cited in a lawsuit. "This is a mistake. This can't be true. They must have mistaken our product with another. This is not possible. Someone is playing a prank on us." The visit from the process server inextricably triggers the process of denial.

Denial quickly gives way to outrage, the second step. "No way! They can't do this to us! I'll counter-sue and get damages for their allegations to our reputation." Left unchecked, the outrage leads to outright anger, and if it escalates, people are known to behave irrationally, if not destructively. Anger is inevitable whether the allegations have any merit or not and must be expunged before it precipitates reactive behavior that will make the matter worse.

The third stage is bargaining. "Wait a minute. It is not our fault. That was the time I had to substitute another manufacturer's bearing from the one I usually use, but I told them that. I'll call them, and surely, they will accept my explanation."

"Oh no, that was the shipment we had to get out in two days, and we skipped final testing. They pressured us to meet that shipping deadline, so they must share the responsibility. As soon as I explain this to the attorney, he can make this go away."

"I'll bet I can offer them a replacement at no charge, and they will drop the lawsuit." Bargaining is a convoluted process of creating unreal scenarios that can explain why an error occurred.

Stage four is depression. "This will ruin the company! Our stock will plummet! If this happened, how many other ticking time bombs are out there waiting to explode on us?" This stage makes it difficult for a consulting client to see that there may be solutions that do not lead to total devastation. This is when defendants are least rational about the steps to prove their innocence or minimize their exposure.

There is no solution until there is acceptance, which is the last stage. After acceptance, we can perform triage to minimize the damage while we implement long-term

DOI: 10.1201/9781003535621-9

treatments for the disease. Unfortunately, all too many organizations never get to this stage. They mire themselves in one of the first four stages and refuse to deal with the issues objectively.

If you can identify which stage the defendant is in, the mapping strategy becomes more exact, and outcomes are more certain. For defense counsel, if the client is not in the acceptance stage, it would be wise to get the insurance carrier to offer the limits of coverage with an agreement for non-disclosure of the terms of a settlement. For the plaintiff, identifying that a defendant is still within one of the first four stages of grief is critical. Your client may blow the case if they are still in an emotional state when deposed or testifying.

As a businessperson, you might want to schedule a Kübler-Ross seminar into training sessions. It may seem superfluous, but being able to recover from even minor tragedies quickly and objectively can be a profitable strategic business tactic.

7.2   SECTION SUMMARY

In this Section, I have journeyed from outer space to the pits of depression and grief. I have examined businesses as living organisms and performed complete physical exams, given clean bills to the healthy, and diagnosed cancer in others.

I have built a viable foundation for understanding how to examine businesses forensically and pathologically for objective evidence of appropriate or inappropriate standard of care. I have given examples of how vision, mission, and values are as important to product and service quality, reliability, and safety as the integrity of the components and the artisanship of the workers.

I explored the world of quality management and the world of standards such as ISO 9000 and created a foundation for our later incorporation of industry standards into building a winning case strategy. The most astute business leaders are seldom aware that risk management is a proactive process that saves money and enhances customer satisfaction. I have created a checklist of symptoms exhibited by diseased organizations and how the maladies manifest themselves in products liability.

I have borrowed concepts such as the Kairos Moment and used oxymorons such as enterprise entropy and paradigm paralysis to dramatize how narrow case research may be in finding the root cause of a tragic accident.

If this Section has been effective in its intended goal, litigators should begin seeing cases from a more holistic perspective of a living organism that has contracted a disease. And business leaders should be more aware that they may be living in a cocoon, unaware of the dangers lurking within their own environments.

Section 3

The Limitations of Risk Management

8 Risk Management

8.1 DEFINITIONS

The *Oxford Dictionary* defines risk as a situation involving exposure to danger. According to the Risk Management Institute, risk management involves understanding, analyzing, and addressing risk to make sure organizations achieve their objectives.

ISO/TC 176/SC2/N1284 guidance document for ISO9001:2015 suggests that risk-based thinking is intended to establish a systematic approach to considering risk rather than treating "prevention" as a separate component of a quality management system. It goes on to state that risk-based thinking is something we all do automatically in everyday life.

ISO 31000:2018 states that the purpose of risk management is the creation and protection of value. In the eight principles outlined in the standard, one tenet includes risk management, which anticipates, detects, acknowledges, and responds to changes.

For any topic, Wikipedia has a page of collaborative knowledge that many make their first stop in research, and some even accept as an authoritative pronouncement. While it is often helpful to research ideas there, remember they may be only opinions. Wikipedia defines risk management as the identification, evaluation, and prioritization of risks, followed by coordinated and economical application of resources to minimize, monitor, and control the probability or impact of unfortunate events or to maximize opportunities.

"Financial risk management" is the protection of economic value in a firm by using financial instruments to manage risk exposure. These include operational risk, credit and market risk, and many others. Addressing financial risk is always punitive, and it is accomplished by auditing after the fact while expecting to find nonconformities. This process is extremely expensive and time-consuming, and there are those who are constantly looking for ways to circumvent financial risk-prevention tools. However, it's worth noting here that in process risk, I don't expect to find individuals creating workarounds for reasonable risk avoidance as is found in financial risk auditing.

"Operational risk management" or ORM is a continual cyclic process. It includes risk assessment, risk decision-making, and implementing risk controls that result in acceptance, mitigation, or avoidance of risk. ORM is defined as the oversight of operational risk, including the risk of loss. This can result from inadequate or failed internal processes and systems, human factors, or external events.

"Risk assessment," or RA, determines mishaps, their likelihood and consequences, and the tolerances for such events. The results of this process may be expressed in a quantitative or qualitative fashion. Risk assessment is an inherent part of a broader risk-management strategy to help reduce risk-related consequences.

DOI: 10.1201/9781003535621-11

Risk management in healthcare comprises clinical and administrative systems, processes, and reports. These are used to detect, monitor, assess, mitigate, and prevent risks. The Joint Commission is the certifying agent for healthcare facilities that want to accept insurance claims. Their audits and findings are very prescriptive and punitive.

Insurance risk management has focused on "pure risks," in other words, loss by fortuitous or accidental means. Insurance risk is based on actuarial tables of calculated risk. This is an enormous industry that calculates not only risk but also the ravages of natural and man-made disasters that can bankrupt an insurance company in only a few days.

There are many more categories of risk management. They include:

- Customer Credit Risk Management
- Industry-Specific Strategy
- Elimination of Contract Risk
- Compliance Risks
- Safety Risks
- Information Security Risk
- Market Risk

Which definition of risk management is closest to your personal understanding of the term? Which, or how many, apply to your organization? What typically happens after assessment and categorization?

Please take time to begin a first-cut definition of your interpretation of risk management in your organization. This will be very helpful as I begin discussing formal assessments. *None of the definitions I've covered will create a robust business management system that is focused only on risk avoidance.*

In typical risk management, once traditional risks have been identified and assessed, all techniques to manage the risk fall into one or more of these four major categories:

- Ambivalence (eliminate, withdraw from, or not become involved)
- Reduction (optimize—mitigate)
- Sharing (transfer—outsource or insure)
- Retention (accept and budget)

None of these are cures; they just mask or defer symptoms that delay the use of root-cause analysis.

8.2   YOUR RISK

These definitions of "risk" run from something we avoid automatically in our daily lives to avoiding danger in business, protecting value, adhering to regulations, and avoiding a lawsuit. But the key questions for us are:

- How do enlightened business leaders decide which definition is appropriate for their organization?
- How do quality professionals provide guidance and tools for risk avoidance in those organizations?
- How do organizations immunize themselves from civil litigation?

In recent years, the quality profession has embraced risk management as the evolution of preventive action. Unfortunately, the same term, "risk management," describes an entire series of industries that focus on statutory and regulatory compliance and monetizing the cost of risk. As professionals, we currently focus on compliance, not excellence. Because of this, it is now increasingly recognized that some of the tenets of quality management are fundamentally flawed, such as focusing on minimizing defects instead of ensuring a critical defect never reaches a customer.

There's one more critical definition of risk that can have devastating results. In legal parlance, "risk" is the potential danger that threatens to harm or destroy an object, event, or person. It's the basis for products liability and organizational negligence litigation. Our work as expert witnesses uses quality standards and auditing techniques to prove an acceptable or negligent standard of care by a defendant company.

In the definition of "risk-based thinking" in ISO 9001:2015, risk management is something we all do in everyday life. This is an oxymoron—what we really need to do is learn how to think strategically. Reading any newspaper or social media will reveal that strategic thinking is woefully lacking in society.

Let's look at the definition of strategic thinking: An intentional and rational thought process that focuses on the analysis of critical factors and variables that will influence the long-term success of a business, a team, or an individual.

Does that sound like our behavior when solving a problem? Or do most of us rely on gut instinct rather than a logical process? As quality professionals, relying primarily on our instincts could put us out of business.

And now we arrive at the critical point in our discussion of risk. In the QM Courseware, we create new models of business management, risk avoidance, and foreseeable risk. We created a cultural imperative to make risk avoidance the cornerstone of business. The roles of quality professionals will expand from quality management to an enterprise-wide system of proactive business management.

8.3   LEGAL RISK

In legal parlance, "risk" is the basis for products liability and organizational negligence litigation. My work as an expert witness uses quality standards auditing techniques to prove an acceptable or negligent standard of care by a defendant company. The legal definition of risk is the potential danger that threatens to harm or destroy an object, event, or person. Examples of legal risk are covered in detail in Section 5.

Section 4

The Limitations of
Process Management

9 Process Management

9.1 MANAGING PROCESSES

The fathers of quality learned early on that to control outgoing quality, it was mandatory to manage the processes that lead to the final product or service. Everything we do in life and business is a process. It has an input, an activity, and a measurable outcome. Brushing your teeth is a process. It begins with a toothbrush and toothpaste being supplied from an outside source. Brushing, rinsing, and flossing are the activities. The output is measurable as you maintain good oral health over a long period of time.

In manufacturing, a process might be receiving a kit of parts from the stockroom, assembling the product, testing the product, and returning it to inventory as either a completed assembly or a work in progress. The drawings tell us the components we need. The documentation provides instructions for assembly. The operator is competent in the required skills. The outcome is often inspected by a third party for conformance.

Quality professionals have worked diligently on tools to manage processes with the goal of minimizing defects. Some methods are the result of trial and error. An associate told me recently that he manufactured a product years ago that must produce outcomes within a certain tolerance. They used precision resistors to achieve that outcome, but they were never more than 94% successful because the precision resistors were not always the tolerance as labeled. They managed the process by having an acceptable defect level that really was not acceptable at all. In Section 7, you will be exposed to my breakthrough methodology for risk avoidance.

Conversely, Motorola created a much more scientific approach to process management. They established that Six Sigma level of quality was their acceptable defect level in their business model. It evolved when they attempted to lower the defect rate in their Quasar TV products. When they started manufacturing cell phones, Six Sigma was the tool they used to be able to assemble a phone and package it without a final test step. For the cell phone market, those few defective products were an acceptable business outcome. A defective cell phone was not considered a critical defect.

When Six Sigma went from a process management tool to a cult, implementers turned it into a one-size-fits-all for manufacturing companies. The Jack Welch success story at GE took the quality world by storm. Unfortunately, the purveyors did not mention that Jack Welch owned GE Credit, which financed this grand experiment. Most companies that have implemented it have discovered that the cost-benefit ratio is not just there for organizations that do not do mass production.

The other critical issue with current process management models is that zero critical defects reaching the customer is the only acceptable defect rate. The most striking examples are pharma and medical devices. Prescription and OTC meds can contain ANY defective ingredients. Pharma companies must also meet the rigors of the FDA,

DOI: 10.1201/9781003535621-13

so there are no acceptable levels of defects. I worked with a big pharma distribution center for a time. In 12 years, they had never shipped a product that was incorrect or out of shelf life. Never. That is the degree of diligence that pharma must have. When we teach their model to others, the senior managers always reply that it is not cost-effective to ship products that are 100% defect-free. You will find in Section 7 how to debunk this myth and put ineffective process management in the rear-view mirror.

9.2   THE PHENOMENA OF TRIBAL KNOWLEDGE

I am truly blessed to have worked with more than 700 companies over my career. I have seen the good, the bad, and the ugly. I use this knowledge to help others avoid the pitfalls I have seen across many industries. That's why I teach risk avoidance instead of risk management. In Section 2, I deal with the symptoms of pathological organizations, their signs, and their disease progression. Nothing is more disastrous to the health of an organization than the existence of sub-groups and operation by tribal knowledge.

Most of us are victims of silos within an organization. That is, a particular department or discipline allows the members to believe that the entire organization revolves around them and that they are the critical path that everyone should respect. An example might be an inventory control department. They receive all the goods and work in progress, and they supply the lifeblood of their products to manufacturing. This can lead to the head of the group creating a culture that purchasing only exists to supply them and manufacturing only exists to turn their supplies into finished goods. The reality is that each is a process that creates the whole. Purchasing procures the raw materials, manufacturing assembles and tests the products, inventory controls stores and controls finished goods, and shipping affects delivery to the customers.

Each is a standalone process, not a silo disconnected from the world. Within those silos are the keepers of tribal knowledge. As organizations evolve and mature, the issues of business growth typically keep the founding members from documenting procedures. Even more often, entrepreneurs are so focused on initial success that they are oblivious as to how they were able to assemble a group of like-minded individuals to move the organization from start-up disaster to start-up success. The need to create a cultural vision, mission, and values is always unknown until employees are added. Even then, onboarding new people is usually a matter of finding people who look like they will fit in and have needed skills rather than those who came on expecting to become part of a successful team.

These new members may very well become assimilated into the undefined business culture, but the longer they go without codified mission, vision, and values, the more likely the company will run on tribal knowledge that has no chance of long-term success.

Early tribal knowledge is a methodology handed down from more senior members who have pragmatically created informal procedures and expectations. These teachings have worked through the start-up phase, and there is little awareness that training new people requires documented procedures, roles, and responsibilities that are validated and approved by senior management. If a company is going to grow and prosper, it is mandatory that operational procedures be documented, used as training aides, and changed by a formal revision procedure, not because someone has wild hair to try something new.

Later in company development, undocumented procedures tend to allow silos to be formed and informal leaders to take over as self-proclaimed experts. This pragmatic growth is again a formula for disaster. It can also happen when new middle management is hired, and they attempt to influence how procedures are followed, attempting to change the culture to how they did it in the past. Ambitious employees can and are of value when they make recommendations, but the culture of a company can only be successfully modified by senior management who develops the mission, vision, and values. It is very difficult to dismantle established silos. Undoing perceived power can be very painful, but it is essential for the overall health of the organization. In one company, the lead salesperson controlled what prospects the rest of the sales staff received. He was always on top of the sales leaderboard, but the sales volume was not acceptable to the senior staff. I made the case that the commission structure needed to be modified so that an individual's commission would not only be based on their sales but also on the sales of the entire department. The sales leader saw the handwriting on the wall. He started mentoring the other salespeople and sharing the leads. Everybody won in that scenario. As the adage states, all boats rise on the high tide.

9.3 INTERNAL AUDITS AND PROCESS CONTROL

A common definition for the globally accepted scheme of periodic compliance audits is a systematic, independent, and documented process for obtaining objective evidence and evaluating it objectively to determine the extent to which the audit criteria are fulfilled. In the case of an audit of an ISO 9001:2015 QMS, the audit criteria are the Standard, internal operational procedures, and subordinate QMS documentation.

As defined, classic QMS internal auditing is restricted in scope to the QMS and does not include the overall business operations. Third-party QMS auditing is even more prescriptive since the auditing agency typically provides a written scope of the audit and must stay within that scope.

The output of these audits is most often simply an audit report. This report typically cites what was observed and compares to the governing procedures. The audit checklist used to create the audit report is often filed as background information and not included in the report. Within the report are statements regarding the conformance of a process compared to its stated objectives. Unless a mature set of measurable metrics for key internal processes (e.g., Engineering, Supplier Control, Purchasing, etc.) have been deployed, these statements are typically more opinion than forensic evidence.

The auditor is then free to make comments on the effectiveness of the process against their own training and experiences. The auditor can add any observations for corrective action or improvement opportunities. These are typically experiential and not compliance-oriented.

Finally, the audit report opines whether the QMS is compliant with the Standard as a whole or whichever Clauses are being audited. This conclusion is often mitigated with the successful resolution of minor or major findings. The response to the

TABLE 9.1
Typical ISO 9001:2015 Audit Checklist[1]

#	ISO CLAUSE	Audit Question	Compliant?	Audit Evidence
1	6.1	Has the company determined the risks associated with the manufacturing process?		1
2	6.1	Is there any evidence of the effectiveness of the actions taken to address the risks in the production process?		2
3	6.1	Does the company identify and implement opportunities in the production process?		3
4	7.1	If there is a maintenance procedure for production equipment and tools, has it been followed?		4
5	7.1	Are there any records for the maintenance of the production equipment and production tools?		5
6	7.1	Are gauge calibrations up to date?		6
7	7.1	Are the work environment and equipment appropriate for the job safe and clean?		7
8	7.2	Has the worker been appropriately trained on the job conducted? Are there records for this training?		8
9	7.2	Are suitable methods used to verifying training effectiveness for the employees involved in the manufacturing process?		9
10	7.5	Is the necessary documentation available and suitable for its use in production? This includes work instructions, job orders, etc.		10
11	7.5	Have workers received the most recent version of that documentation?		11

[1] From Advisera ISO 9001 Academy Free Downloads. The author is a Certified Lead Auditor and Lead Implementer to ISO 9001:2015 by Advisera and is a contributing author to the Advisera Blog.

findings should substantiate a conclusion of compliance or noncompliance. It is rare that a finding goes unresolved or that a QMS is disapproved.

Table 9.1 is a typical checklist for conducting ISO 9001:2015 audits. It is annotated with numbers that correlate to questions about what is and should be included in "Audit Evidence." This narrative moves beyond the typical inclusions of evidentiary documents, procedures, work instructions, job descriptions and the like. This creates the first awareness of the differences between traditional audits and forensic investigations.

The following are challenges to the efficacy of traditional ISO 9001 audit checklists from Table 9.1.

1 How can an auditor evaluate a global question about how the company may or may not have determined risks associated with its manufacturing or services process? This would involve an enterprise-wide risk assessment by trained examiners.

2 Again, this requirement for assessment of evidence risk evaluation is outside the experiential background and credentials of the typical ISO 9001 auditor.

3 What empirical evidence can an auditor evaluate to answer whether the company has identified and implemented opportunities? What is the definition of "opportunities?"

4 Determining if there is a "maintenance procedure" will typically be machine-specific. Finding a sample should be easy. Determining if the procedure is being followed would require step-by-step auditing of the procedure as it is executed. As with #2, this is typically outside the expertise of an ISO 9001 auditor (also see #10).

5 The auditee should be prepared to offer any maintenance records. What does the existence of these records prove about the effectiveness of the process?

6 A sampling of devices to determine if gauge calibrations are up to date is a missed opportunity. Less than 100% of active tool and equipment calibrations being current means the process is broken.

7 What empirical evidence can an auditor evaluate to determine if the work environment and equipment are appropriate? This can only be done at the end of a 100% audit.

8 Training records should be available for each process operator and owner. Determining the appropriateness of the training can only be determined by historical records of performance, not from records of how they were trained.

9 What training does a QMS auditor have to evaluate whether operator training is effective or not? It cannot be determined by a sampling audit or interview. Nor is it discoverable from traditional performance reviews.

10 How does an auditor determine if all the necessary documentation is available and suitable? The auditor would need to be an expert on each process to make this determination.

11 It would be common to check a sample of the available documentation in the organization's revision control logs. Again, such an audit would be inconclusive without checking a larger universe of controlled documents, not a sample. One revision error is too many.

As we were trained in auditing classes, traditional ISO 9001 auditing is a snapshot in time of the processes being audited. The scope of internal and external audits is determined ahead of time, and the process owners are notified of the content and time of the audit. Unfortunately, this often gives the auditee the time to "clean up their act" by presenting an error-free example or rehearsed process that may revert to previously used shortcuts once the audit is completed.

We use non-scientific sampling methods to determine which processes and procedures can be audited during the allotted time. I accept the judgment of the auditor or audit team to determine compliance or nonconformance.

Internal audits typically plan to cover all elements of the QMS over a period of six months or a year. Process variability can sometimes be measured in mere hours or days. External initial certification or transition audits are still sampling audits and do not cover the entire QMS. External surveillance audits may plan to cover the entire QMS once in a three-year certification period.

While it is our cornerstone, traditional QMS auditing is grossly inadequate for determining day-to-day QMS compliance. Its efficacy in discovering foreseeable risk is almost nonexistent. Its ability to identify trends in conformance is limited to an almost accidental discovery of similar shortcomings of the QMS. Its ability to discover opportunities for improvement is determined by the competency and discretion of the audit team.

It is time to move beyond conformance auditing to forensic investigations focused on business process excellence and foreseeable risk. We must differentiate and then ingrain this into a company's culture so that it doesn't end up like the often started and abandoned standalone TQM or Six Sigma approaches.

9.4 COMPARING TYPICAL PROCESS AUDITS TO FORENSIC INVESTIGATIONS

Utilizing ISO 19011:2018 as the architecture for traditional first, second, or third-party audits, comparisons are made for how Forensic Investigations transform audits into opportunities for risk avoidance and business process excellence.[1] Table 9.2 is a comparison of traditional ISO audits to using a Forensic Investigation for your audits. Note that by taking this approach, auditors are fully compliant with ISO but take a deeper dive in their processes and procedures.

9.5 ARE WE ONLY LOOKING FOR CONFORMANCE?

After successfully leading the transition to ISO 9001:2000 at a division of Dell Computer, I wrote a book entitled *Implementing ISO 9001:2000—The Journey from Conformance to Performance*.[2] The book was not a tutorial for conducting a successful transition, but it highlighted the successes Dell had in increasing the performance of their bottom line. That case study is contained elsewhere in this book.

The reason I landed that assignment was my approach to the division VP, who said that implementing the Standard was all about creating quality as a profit center, not an overhead conformance exercise. His quality group had been invested in internal auditing and process control to ensure that they passed their third-party audits. This is the thrust of most ISO 9001 implementations. My question to industry and quality leaders is WHY are they so focused on conformance and compliance instead of making our quest to eliminate nonconformities and create process excellence?

In my experience, the answer is that we are so enmeshed in standards compliance that we have lost sight of the vision to create products and services that are fit for

TABLE 9.2
Comparison of Traditional Audits to Forensic Investigations

Traditional Audit	Forensic Investigation
5.4.1 Roles and responsibilities of the individual(s) managing the audit program	**Principal Investigator, Investigators, Scribe**
5.4.2 Competence of individual(s) managing audit program	Investigator instead of Auditor
5.4.3 Establishing extent of audit program	Plant, department, or logical clauses
5.4.4 Determining audit program resources	Team, needed data/checklists, electronic data access
5.5.1 Implementing audit program	**Train all participants in Forensic Investigations**
5.5.2 Defining the objectives, scope, and criteria for an individual audit	Scope of the investigation is by department, function, or logical consolidation of clauses
5.5.3 Selecting and determining audit methods	**A Process Forensic Investigation (FI)**
5.5.4 Selecting audit team members	Non-conflicted certified investigators
5.5.5 Assigning responsibility for an individual audit to the audit team leader	Lead Investigator plus at least one other Investigator
6.1 Conducting an audit	**Conducting a FI**
6.2.1 Initiating audit	Assemble investigators to plan the FI
6.2.2 Establishing contact with auditee	Notification of Scope **no more than one week prior to the investigation**
6.2.3 Determining feasibility of audit	Principal Investigator working with Process Owner and cell-level workers
6.3.1 Performing review of documented information	Operational Procedures, Work Instructions, **Roles and Responsibilities, Process Matrix, Metrics, KPI's**
6.3.2.1 Risk-based approach to planning	**Risk Avoidance Methodology**
6.3.2.2 Audit planning details	Originate Checklist comprised of Operational Procedures, Work Instructions, **Roles and Responsibilities, Process Matrix, Metrics, KPIs, Results from previous audits**
6.3.3 Assigning work to audit team	
6.3.4 Preparing documented information for audit	
6.4.1 Conducting audit activities	A checklist of the planned outcomes of each process
6.4.2 Assigning roles and responsibilities of guides and observers	Determined by the Principal Investigator and Process Owner
6.4.3 Conducting opening meeting	Layout the construct of a FI
6.4.4 Communicating during audit	Principal Investigator moderates communications
6.4.7 Collecting and verifying information	**Evidence-based evaluation of all operations and roles and responsibilities**
6.4.8 Generating audit findings	**Fact-based recommendations for immediate corrective action and process improvements as warranted and generating process excellence recommendations**

(Continued)

TABLE 9.2 *(Continued)*
Comparison of Traditional Audits to Forensic Investigations

Traditional Audit	Forensic Investigation
6.4.9 Determining audit conclusions	**Specific process improvement recommendations with anticipated enhancements in metrics and KPIs**
6.4.9.2 Content of audit conclusions	**Includes Foreseeable Risk Assessments**
6.4.10 Conducting closing meeting	Investigation results compared to anticipated outcomes
6.5 Preparing and distributing audit report	As specified in the appropriate procedure
6.6 Completing audit	Scheduling corrective action assessment reviews, following up on action items, **reporting foreseeable risks, and business process excellence recommendations to senior management**

their intended use. Instead, we are led to believe that if we are compliant with the clauses of ISO 9001, we will have a well-functioning organization and will have a greater customer base because we are certified. Conformance to the Standard means that on any given audit day, one or more processes were deemed compliant with the requirements of the QMS.

In Section 7, I will make the case that all conventional methods to ensure conformance and compliance are inadequate for running a successful business in the 21st century.

NOTES

1 Reference: Conducting Forensic Investigations—The New Gold Standard in Process Auditing©
2 *Implementing ISO 9001:2000; The Journey from Conformance to Performance*, Prentice Hall (2002) ISBN 0-13-061909-4.

Section 5

The Paradigm Shift

10 A Moment of Clarity

10.1 THE CALL FROM AN ATTORNEY

One fine day, an attorney called me, representing the defendant in a lawsuit. The attorney asked if I would look at the case from a "quality control perspective." Later that week, a delivery truck arrived with half a dozen bankers' boxes full of documents.

The defendant was Globe Metallurgical Inc., Beverly, Ohio plant. They manufacture metallurgical and chemical-grade silicon metal and silicon-based alloys. At the time, the Beverly plant produced silicon metal used by the cosmetics industry.

The plaintiff was Titanium Metals (Timet), a large multinational aerospace company that manufactures titanium parts for aircraft, such as jet turbine blades and rotors for helicopters. Timet purchased silicon metal from Globe from 1989 to 1997. During that time, there were 13 purchase orders for 500 pounds of silicon metal each. The batch that led to the lawsuit was for 5,000 pounds.

In the timeframe that led to the lawsuit, Timet was certified to ISO 9001:1994. Globe was certified to ISO 9002:1994. ISO 9002 was part of the original subset of ISO 9000 and was primarily chosen by companies that did not do fundamental design. Globe also won the Malcolm Baldrige National Quality Award in 1988 and the Shingo Prize for manufacturing excellence in 1989.

The litigation occurred in 2003 and 2004. The data included the various legal filings that went back and forth, specifications, purchase orders, and test reports. Each party can request "discovery" from the opposing side to gather any specific documents that they feel are pertinent to the case. I immediately requested the quality manual and any operational procedures on supplier approval, incoming inspection procedures, and the defendant's supplier history.

Only a miniscule amount of silicon metal is needed to forge titanium. Globe kept the product in open bins. For the batch that was purchased which led to the litigation, a light bulb had blown up overhead, and the glass and tungsten filament fell into the bin. Tungsten is known as one of the toughest elements found in nature. It is super dense and almost impossible to melt. It is very resistant to corrosion and has the highest melting point and highest tensile strength of any element. The plaintiff alleged that the supplier shipped the lot in question without testing it for contamination.

For many years, Timet kept its inventory and purchase records on Kardex cards. Each supplier had a card with manually entered information on what was purchased, quantities, and batch numbers. There was also a place for someone from the quality department to indicate annually that the supplier was approved. There was no such approval on the Globe purchase card.

As I was sorting through the data, it occurred to me that I was doing a QMS audit without having to visit the company. There was more than enough documentation to construct the chain of events of how pieces of tungsten were forged into turbine blades and helicopter rotors and how it wasn't discovered until the customers received the final product.

In my report, I drafted a timeline of each batch received, a copy of the receiving report, and a copy of the certificate of conformance. The purchase order stated that the certificate indicated the purity of the silicon metal and the percentages of iron and calcium in the report but nothing else. The certificates were all initialed and dated, indicating the received batch was acceptable.

The next step was to review the supplier approval process and find supporting data for approval and subsequent reapproval. In the documentation, I found a one-page questionnaire dated October 1991 that first put Globe on the approved supplier list. Every few years, I found letters to the defendant notifying it of the plaintiff's intent to conduct an onsite audit. Every time, the plaintiff answered that they did not do enough business with the plaintiff to warrant hosting a facility audit. Somehow, these letters from Globe declining onsite inspection kept the supplier on the plaintiff's approved list for more than ten years. Finally, Globe responded by refusing to sell silicon metal to Timet.

Also, Globe did not agree to be compliant with Timet's Raw Materials Specification 66. Section 2 of that specification required the batches to be free of inclusions such as tungsten, but Globe had no X-ray or fluoroscopic equipment to perform such tests. Globe's certificates of conformance included only a data sheet accepted by Timet throughout their time as supplier and customer.

The final step in my investigation revealed that Timet never tested the raw material received from the defendant. Thus, the batch of contaminated material made it through the entire fabrication, inspection, and test process, with the shards of tungsten forged into the parts going undetected. Since tungsten has a much higher melting point than used in the titanium-forging process, these shards would eventually cause a fracture that would break the turbine and rotor blades. Fortunately, the plaintiff's customers did inspect them and found the defects before the parts were installed. Timet continually released Globe's materials to production with only the data on Globe's certification. Timet never explained why they did not do fluoroscopic testing at the incoming inspection.

In Globe's deposition, it was revealed that its silicon metal was mostly used for cosmetics, and the silicon was never meant to be used in aerospace. My report, deposition, and testimony in court were a storyboard of how Timet had systematically ignored its procedures for supplier approval, incoming testing, and final testing. The judge found Globe not guilty.

After my courtroom appearance and verdict, I came to the realization that I had utilized my skills as a QMS auditor investigating the issues of the case and then used root cause analysis to reach my conclusions and opinions. It's not often in civil and criminal cases that irrefutable evidence is produced by either side. In products liability and organizational negligence cases, however, it is a slam dunk when you can document that the company did not follow its own procedures.

Over the last two decades and 42 more products liability cases, I coined my methodology Forensic Business Pathology® (FBP) and registered it as my trademarked title for evolving traditional management systems auditing into what I now call Forensic Investigations©. In the following chapters, I will create the foundation of how I have pioneered the future of the quality professions from the lessons learned from products liability and organizational negligence lawsuits.

10.2   THE PATH FROM WARRANTY TO LITIGATION

The data presented in this Section is a compilation of professional experiences and case studies in local, state, and federal courts in the Continental United States. It extracts learning moments from products liability and organizational negligence litigation, providing experiential and anecdotal information for attorneys, judges, and other professionals. The data presented is a collection of tools to assist in forming systematic and tactical strategies. It is not legal advice or counsel. The case studies provide anecdotal information. They are not quotable as case law.

I am an expert who offers consultation and opinions within the usual constraints of a non-testifying expert. That is, the reader has not contracted with the author for specific services and may not quote the author in the context of a specific case or circumstance.

I hold myself harmless from any negative outcomes experienced using any information in this Section. This is a compendium of anecdotal and experiential data synthesized into a cookbook for attorneys and business professionals, not a clinical or reference text.

The preceding paragraphs represent the intent of starting the Section with a discussion about products liability and organizational negligence in the United States. Each statement is factual. Each statement should be self-evident. A reasonable person would not base the outcome of a trial on a quote from this Section. Each of my disclaimers should prevail in a legal action, but none of these statements is incontrovertible. Ergo, we have the paradox of the American legal system.

Many of the readers have no doubt witnessed the most compelling arguments and evidence rendered impotent by a judge or jury either in real life or in the media. I have all seen plaintiff and defense litigators turn rational and accurate testimony into a nightmare of unresolvable riddles. I have observed incontrovertible forensic evidence discredited by procedural issues. I have read opinions that obfuscated the most clearly written statutes. Most frustrating, I have seen emotion trump fact in the face of high drama.

I rejoice that I live within the protection of the most effective legal system in the free world while I toil under the fact that justice is not always "fair." A jury may award a multi-million-dollar settlement to a migrant worker in South Texas, while another jury in California may render a verdict in favor of a celebrity who appears to get away with murder. Present day, our system of jurisprudence has become so polluted with politics that we can often not tell the good guys from the bad. The founding fathers would be turning over in their graves at what has happened to our Constitution.

More than a conundrum or an anachronism, the American system of jurisprudence is perpetual motion incarnate. It is a chess game of strategy where a worthy opponent can unexpectedly foil classical moves with obtuse strategy. The most well-prepared game plan can crumble with a subtle variation on the traditional play or the introduction of obtuse logic.

It is the world series of Texas Hold 'em, where hundreds of games are played concurrently but where no two games are the same and winning strategies at one table may be ineffective against a skillful bluff at another. The poker metaphor is even more pertinent because Texas Hold 'em is a game of concise strategy and numbers combined with skillful use of illusion and deceit to gain a winning outcome. I will deal with these paradoxical issues in more detail in later chapters, and I will reveal how to win the game without dark glasses and baseless bluffing.

As I begin our journey into exploring new strategies in products liability and organizational negligence tort, there is a corollary to the disclaimer that the Section is applicable only to the American system of justice, which is critical to present at this early juncture. The Section is about American companies and the incredible success stories they precipitate. I also create some of the most egregious examples of immoral and unethical behavior. Our free society and entrepreneurial culture can produce the most extraordinary medical breakthroughs and the most cunning Ponzi schemes, both under the creative leadership of similarly gifted geniuses. The gamete of business models also ranges from the most dedicated to excellence, reliability, and safety to those that use an "acceptable kill level" model in deciding cost over quality. Besides overt immoral and unethical behavior, there is a systemic scenario that may be the most likely to produce the most defective products and harmful services that lead to death, injury, and litigation in our Country. As we discussed earlier, I have labeled the phenomenon Enterprise Entropy.

Corporate leaders are unconsciously abdicating their standards of care for higher profits. The pressures of customer and production issues can cause otherwise conscientious managers to become oblivious to decaying levels of service. Awareness of the potential harm of a product or service is too often invisible until a process server delivers a summons for civil litigation to an unsuspecting CEO. Again, I address this topic in detail, both in the context of legal strategy and as a prophylaxis for business leaders to avoid lawsuits later in the Section.

To reiterate, the central theme of this Section proposes that:

- Personal injury and death resulting from product-related incidents are the result of consumers using marginally safe products incorrectly.
- Product failures are more often the result of processes out of control and companies abdicating their stewardship and accountability than from poor product design.

While the theme is not necessarily unique to the American culture, this Section will stay focused on how the peculiar complexities and paradoxes of our society are continually challenging our legal system as it exists and as it evolves. I will also focus on the common behavior among American companies that is pathological and damaging. Along with awareness of liability forensics for members of the legal community, it is my hope that enlightened business leaders will also embrace the contents of this Section as a model for liability "avoidance" in their organizations.

11 Case Studies

11.1 MOTIVATION

The motivation to write this book is partly my passion for excellence precipitated by my intimate involvement in the early days of the space program. It has been amplified more recently by my extensive experience as an expert witness in products liability and organizational negligence. My last book, Foreseeable Risk, is a desk reference for business leaders who are astute enough to embrace the tenets of avoiding liability and doing no harm. It is also a guide for those who have been stricken with lawsuits that have crippled their organizations and turned their personal lives into living nightmares.

This chapter is dedicated to giving the reader the impetus to continue reading this book and to take to heart the imperatives of what I call the Apollo Business Model. Whether you are just challenged by the complexity of business dynamics or are looking to turn around a foundering organization, I have remarkable solutions to overwhelming problems in the pages ahead. I also hope you will join us in the mission to end the trend toward mediocrity, which has become so pervasive in the first decade of the 21st Century.

The following headlines and summaries are from actual lawsuits in which I have given testimony. The names are withheld because of confidentiality agreements and to protect the guilty. The Timet case I cited earlier is public record.

11.2 FAMILY PERISHES IN HOUSE FIRE

Investigators found that the electrical receptacle for a window air conditioner was the origin of a fire that led to the demise of a family as they slept. Forensic science concluded that the outlet was defective and unsafe for its intended use. Investigation of the manufacturer discovered that the outlet was assembled offshore in a facility controlled by a US holding company. The compelling evidence of negligence from discovery led to the judge allowing the plaintiff's representatives to conduct an inspection at the offshore factory.

There, we discovered fabrication, assembly, and inspection processes that were grossly substandard. The production equipment had been transported from the manufacturing facility of the previous owner in the US and was being operated by workers who had no training and had no idea what they were building.

While the product was UL listed for safety and was under ongoing prescribed surveillance in the US, the requirements of the UL listing had been grossly ignored and never under adequate inspection at the offshore facility. Changes had been made to the product that were not discoverable without forensic testing and there was no testing done on the product after assembly.

DOI: 10.1201/9781003535621-16

Case Study #11.2

This electrical outlet was installed on a board in a forensic laboratory. For this lawsuit, it replicated the outlet that was installed at the plaintiff's home that caused a fatal fire.

Case Study #11.3

This photo depicts an electric space heater in the testing laboratory of the company named as the defendant in this case. It spontaneously caught fire because of defective parts.

The survivors received cash in a settlement that has been sealed so that the public will never be aware of how dangerous the product is. Yes, a version of this product is still for sale all over the US.

11.3   ANOTHER FAMILY PERISHES IN HOUSE FIRE

Investigators have traced the cause of a house fire that killed several people to a portable space heater. A forensic investigation confirmed that electrical defects in the heater were, in fact, the cause of the ignition of the blaze that consumed the residents in their sleep.

The space heater had the brand name of an American company. Attempting to trace the actual location where it was manufactured was obfuscated by a quagmire of holding companies, importers, and multiple factories that made similar products.

In their advertising, the brand company made many claims of extreme detail to quality and safety. When their manufacturer was in Asia, it was discovered that the requirements of the UL safety listing for that product had been compromised by using inferior plastics that were not able to contain sparks that led to the combustion of the housing and the surrounding furnishings and structure.

The survivors received cash in a settlement that has been sealed so that the public will never be aware of how dangerous the product is. Yes, a version of this product is still for sale all over the US.

11.4   MAN DIES IN HOUSE FIRE

Officials discovered an elderly gentleman dead in his bedroom after a house fire destroyed most of his home. The origin of the fire was traced to an oxygen concentrator used by patients who require supplemental oxygen.

Forensic evidence pointed to the machine as the source of the pure oxygen that allowed the uncontrolled acceleration of the fire. Detailed analysis uncovered that the deceased had been smoking a cigarette that came in contact with the plastic hose carrying the oxygen across the room.

While it might be opined that the man's cigarette smoking in bed while using oxygen was certainly a bad idea, further investigation of the machine's manufacturer disclosed that this was not the first time they had encountered this "issue." In fact, they had been involved in previous litigation, and they were made aware that a very inexpensive check valve could stop the flame from propagating back to the machine that was generating the oxygen.

While the victim may have acted inappropriately, the manufacturer was aware that there was a way to prevent such misuse but failed to exercise and provide an appropriate standard of care.

The survivors received cash in a settlement that has been sealed so that the public will never be aware of how dangerous the product is. Yes, a version of this product is still for sale all over the US.

11.5   AND AGAIN, A FAMILY PERISHES IN HOUSE FIRE

Investigators have traced the cause of a house fire that killed several people to a box fan. A forensic investigation confirmed that electrical defects in the fan were, in fact, the cause of the ignition of the blaze that consumed the residents in their sleep.

Detailed investigation revealed that the defect was the result of a poorly manufactured motor and switch assembly that caused the sparking that led to the ignition of the fire. The fan was constructed in the U.S. by a company that was in the process of offshoring the manufacturing and closing of their U.S. plant.

The compelling evidence of negligence from discovery led to the judge allowing the plaintiff's representatives to conduct an inspection at the factory. That audit produced evidence that the motor and switch assembly had an alarmingly high failure

Case Study #11.5

This photo is of a common box fan that caught on fire because of a defective switch.

rate as they were received from the Asian manufacturer. Examination of the purchasing documents showed that the manufacturer anticipated a high failure rate and was given a failure allowance in the agreement with the supplier.

I also observed an extraordinary number of fans that were removed from the production line when they failed the final inspection and test. The production line was routinely shut down each afternoon to rework defective fans.

The survivors received cash in a settlement that has been sealed so that the public will never be aware of how dangerous the product is. It is unknown if the offshore version of this fan is for sale at U.S. retailers.

Even to the untrained eye, a pattern of manufacturers exhibiting an unacceptable standard of care in placing defective electrical products into retail outlets in the U.S. is alarming, if not shocking (pun intended). But wait, I am not through with citing case examples that are just from my personal experiences in less than a decade of expert witness work.

11.6   YOUNGSTER ELECTROCUTED WHILE USING A BATTERY CHARGER

A young man was found nonresponsive by first responders in the backyard of his family's home. The youth was barefoot and attempting to connect a battery charger to his go-kart when he was electrocuted. It is theorized that the first jolt knocked the boy to the ground, and the metal handle of the device fell across his chest, causing the fatal electrical current flow.

A forensic investigation discovered that a poorly positioned wire bundle had come in contact with the cooling fan mechanism inside the charger. Over time, the

Case Study #11.6

This photo is of an old-style battery charger that was defective and electrocuted a young boy.

insulation became frayed, and eventually, lethal line voltage was conducted to the frame of the device.

From the forensic investigators' findings, I was able to prove that the manufacturer did not exercise an appropriate standard of care in designing and manufacturing the device. The survivors received cash in a settlement that has been sealed so that the public will never be aware of how dangerous the product is. A version of this product is still for sale all over the U.S. It is now made of plastic instead of metal.

11.7   FAMILY PERISHES IN HOUSE FIRE (ARE YOU GETTING THE DRIFT?)

Investigators traced the origin of a house fire to a "boom box" in the living room of the home where members of a family perished from smoke inhalation. The CD player and radio were part of an entertainment center that was turned off during the night of the incident. Forensic investigation found that the plastic case of the device had ignited and dripped flaming plastic onto the surrounding furnishings and carpet.

Most such entertainment devices are actually "on," even when they are turned off. This is to allow remote control devices to be used to activate the main circuitry. Our investigation determined that the power supply, which was on continually, had defects in design and manufacture that allowed overheating of electrical connections to the point that they could cause ignition. Also, the plastic housing used by the Asian manufacturer was not flame retardant as required by their UL listing.

The survivors received cash in a settlement that has been sealed so that the public will never be aware of how dangerous the product is. Yes, a version of this product is still for sale all over the U.S.

11.8   OILFIELD WORKER PERISHES IN WELL-HEAD ACCIDENT

Company personnel discovered the charred remains of a maintenance technician beside a flaming wellhead. The worker was performing routine maintenance duties on an old production site when the incident took place.

A forensic investigation traced the source of the ignition to a faulty valve that had been installed on the pipe where the well came out of the ground. Our investigation discovered that the valve used was not the one specified for use by the owners of the well. In fact, the chain of custody showed that a distributor had substituted a valve meant only for use in water wells had been substituted for the one that would have been certified for use in high-pressure oil and gas wells.

Even though the threads of the valve were significantly different from the mating pipe, a previous maintenance technician had somehow managed to cross-thread the valve in place, creating the opportunity for the valve to blow out violently, contributing to the fire ignition.

Case Study #11.8

This photo is of a producing oil well after it exploded, causing the death of a service technician.

The survivors received cash in a settlement that has been sealed so that the public will never be aware of how dangerous the product is.

11.9 CHILD DIES FROM ADMINISTRATION OF INCORRECT DRUG

Hospital personnel were unable to revive a young girl who was administered an incorrect intravenous medication. The youngster was admitted with breathing problems, and an IV was ordered with a medicine that was designed to relieve her congestion. Forensic investigation revealed that the hospital pharmacy had dispensed a medication that had a very similar name but was formulated to induce coma in seriously injured adult patients.

Our investigation determined that the automated prescription dispensing machine did not have appropriate safeguards to prevent erroneously dispensing drugs with similar names. Human error contributed to the incorrect product being admitted to the patient, but the machine manufacturer shared the liability for the fatality.

The survivors received cash in a settlement that has been sealed so that the public will never be aware of how dangerous the Hospital's procedures are. The machine was redesigned to prevent spelling errors.

At this point, I hope you are as distressed reading these stories as I am having to deal with them over a protracted legal battle. I have seen more photos of dead people who were victims of manufacturing and service companies intentionally or unintentionally, creating a foreseeable risk for the users of the product. I will conclude this horror tale with a few stories that did not cause fatalities but did result in serious injury and/or financial harm.

11.10 WOMAN SERIOUSLY INJURED IN BICYCLE ACCIDENT

A woman sustained serious back and leg injuries when she was thrown off her custom touring bicycle and hit the pavement. The handlebars snapped from their mounting during a routine bike ride.

A forensic investigation determined that a weld had failed on the coupling holding the handlebars, causing them to detach suddenly, leading to the accident. The bike had been custom-built with components from various suppliers selected by the manufacturer.

Our investigations determined that the defective component was manufactured in Asia according to the specifications they provided. Unfortunately, the drawings did not specify how the weld was to be done. In the U.S., The American Society for Testing Materials (ASTM) provides standards for welding integrity that are imposed on welders to ensure reliability and safety. While such standards may exist in other countries, there were no such requirements on the fabrication drawing. Obviously, the welding was substandard.

The woman received cash in a settlement that has been sealed so that the public will never be aware of how dangerous the product is. Yes, a version of this product is still for sale all over the U.S.

11.11   WOMAN SUSTAINS SERIOUS INJURIES AFTER BEING EJECTED FROM A MOVING GOLF CART

First responders treated a woman with substantial lower back injuries on a local golf course. Witnesses said that she was ejected from the cart as it was being driven by another woman from one hole to the next.

A forensic investigation revealed that there were no safety belts, handholds, or other restraints present in the cart. Our investigation revealed that there are typically no such safety devices required for these off-road vehicles.

The woman is petite in stature and could not reach the canopy or front end of the cart to hold on over bumpy and steep terrain. There were no handholds on the seat and the side restraints were too short to grasp.

In discovery, it was revealed that these types of accidents are common, and the manufacturer regularly defended themselves in lawsuits over similar issues. In the deposition, a representative of the manufacturer testified that golfers were only interested in easy ingress and egress from the carts, and that is why the side restraints were much shorter and smaller than one would find in an office chair. He further testified that golfers would never use seat belts and that they would not subject themselves to safety training to operate the carts.

The woman received cash in a settlement that has been sealed so that the public will never be aware of how dangerous the product is. Yes, a version of this product is still used all over the U.S. Next time you get in a golf cart, determine for yourself if you are adequately protected from ejection!

Case Study #11.11

This is a photo taken in this lawsuit. It is a golf cart that rolled over steep terrain and seriously injured one of the passengers.

I have many more examples and anecdotes, but I think I have made the point that there are many hidden dangers in manufactured products and services. While our manufacturers, importers, distributors, and retailers may dismiss these types of problems as "acceptable failure rates," I believe (and it is my experience) that we can build products that are intrinsically safe, do no harm, perform to specification, and be sold at a competitive price.

I am taking manufacturers to task to examine their predisposition for greed in making target sales numbers and looking for viable alternatives to producing products that are substandard and dangerous. OFFSHORING MANUFACTURING IS NOT THE SOLUTION.

The imperatives and business model revealed in this book are a wakeup call for entrepreneurs and leaders to reexamine the calamities and catastrophes such as the collapse of Enron, securities giants, banks, and other iconic organizations and look beyond conventional wisdom and the acceptable kill rate for a new model of moral and ethical business behavior that creates profit, growth, and sustainability based on providing the best and safest products and services at a competitive price, within our own borders.

The 21st Century is unfolding as the era in which we become a third-world political power and a nation that is incapable of providing its own resources, services, and products when we truly have the potential to lead the world once again in all aspects of commerce.

We made up our minds to go to the moon and accomplished that impossible task in seven years with a coalition of businesses from the size of IBM to two-person machine shops. How we did it is waiting for you to learn and adapt to your business and culture before it is too late to save the Republic. Then COVID-19 fundamentally changed our historical business models.

12 The Plaintiff's Stories

12.1 DEFINITION

Plain·tiff (noun), Law. A person who sues in a court (as opposed to the defendant).

The difference between an injured party and a plaintiff is in the subtlety of being just a victim transformed into the need and desire to hold someone else accountable for their injuries and/or losses. Once the transition begins, attorneys weave the gruesome and egregious details of the event(s) into a defensible story that is not only horrible for its aftermath, but there must be compelling evidence that the plaintiff was wronged by individuals or organizations who intentionally, unintentionally or through neglectful behavior caused the tragedy to occur.

Please understand I am not minimizing the disasters I have been witness to, and I never again want to see a photo of a child who perished in a fire. I am laying the groundwork for the "business" of civil litigation.

When the plaintiff's attorneys interview clients, they pose questions based on data gleaned from the injured party's statements. Their memory register stacks are replete with law school training, apprenticeship, the influence of their mentors, case studies they have researched, and their experiences in practice. They receive and compartmentalize each data packet, weaving it into the outline of a story that will become the allegation in the suit and the strategy to overwhelm the defense.

It is a "story" because accidents do not happen in a single time slice. They are typically a serendipitous series of events that, without the influence of any one of the pieces, might not have led to the alleged disastrous outcome. To prevail in mediation or court, the plaintiff's job is to inextricably connect all the pieces, while the defense must discredit one or more of the events. Thus, the plaintiff's story is the most critical path to success (by either side) in the tort process.

Constructing the story outline provides the data for the initial complaint filing. While there are many unknowns, the plaintiff's counsel often names many codefendants if discovery turns up others who are complicit in the disaster.

The remainder of the life of the lawsuit involves choreographing, honing, and preparing the case for negotiation or trial. The plaintiff's attorneys hire experts to sift through the forensic and trace evidence, looking for the smoking gun to "prove" the story. If the defendant engages competent counsel, they will also hire experts to refute the plaintiff's experts. The cause and origin experts are the first to clash over causal issues. Even if there is surveillance video of an arsonist lighting a fire, attorneys may hire experts to determine if there was tampering with the recording. The war between the experts may cause these cases to drag on for years, at great monetary and emotional costs for both sides.

In the coming chapters, the reader will learn that the plaintiff's story can become incontrovertible with the intervention of a new strategy at the opening of the case file and before the story outline is drafted. Decades of experiential data will demonstrate

DOI: 10.1201/9781003535621-17

how to achieve rapid success in these cases by looking beyond the physical evidence of the company that built the product or provided the services that led to the disaster. Snippets from a couple of case studies, discussed later, will prepare us for the journey into the world of Forensic Business Pathology.

In the wellhead fire example, a maintenance technician died in a fire at the wellhead of a producing oil well. The explosion and fire laid waste to most of the pipes and valves. For several years, the forensic experts examined every scrap of evidence and produced conflicting reports, back and forth, as to what caused the fire. The only agreement between the sides was that there must have been natural gas leaking from the wellbore before the explosion. With no compelling evidence to prove the cause, the defense was mounting a case that the technician had parked his truck too close to the wellhead and that he had caused his own demise by not observing accepted safety practices. Their allegation was that, regardless of whether there was natural gas leaking from the well, the ignition for the explosion came from his truck.

An associate referred me to the case by suggesting to the plaintiff's attorney that there might be another approach to establishing liability with reasonable certainty. The attorney was surprised when I declined to look at any of the forensic evidence. Instead, I requested the discovery of the maintenance records for the well for the previous couple of years. After detailed scrutiny, I discovered that another maintenance crew replaced a valve with one that was suitable only for low-pressure water applications. Although the maintenance company allegedly requested the correct valve, the chain of custody from the wholesaler to the distributor to the retailer to the company that installed the valve was negligent because none of them discovered that the valve was noncompliant with American Petroleum Institute (API) standards. Not only was the forensic evidence immaterial to the outcome of the case, but much time and money were spent debating the circumstances of the accident. How the ignition occurred became incidental to the evidence that the valve would have blown out eventually because the pressure present at the wellhead was beyond the limits of the water valve. The question was also raised of how the service technician who replaced the valve managed to get the threads to mate. The two parties reached a settlement before this question could be answered.

In another case, a defense team retained me to look at a file involving an aerosol can that allegedly exploded and injured a worker. Again, the plaintiff and defense spent much time and money examining the subject can and attempting to recreate the exact circumstances of the accident. Both sides presented plausible scenarios, although neither had conclusive arguments. I requested the quality and production records of the manufacturer of the aerosol can, who was one of the named defendants. Close examination of the manufacturing and quality processes revealed that the company had ingeniously installed a super-heated water flume to transfer the completed and charged cans to the stock room. Instead of a conveyor system, the trip along the water flume heated each can well beyond any temperature that might exist in the environment where the explosion happened. Most aerosol cans also have a special seal on their concave bottoms that opens to release the accelerant long before it can explode from heat or trauma. This irrefutable testimony showed that the manufacturer could not have been held negligent in the design and manufacture of the product. Again, the case was settled, and money changed hands even though the plaintiff was proven to be an outright liar.

Attorneys often embrace Forensic Business Pathology out of desperation in the middle of the "story" instead of incorporating it into the initial interview with the plaintiff. The strategy of liability forensics for the manufacturer or service provider will be discussed in more detail in upcoming chapters.

12.2 COMPENSATION

Plaintiffs' attorneys each have their own formula for deciding if a case has merit. The rationale runs the gamut from a desire to advocate for the client to the potential of a large payday. In any event, compensation to the plaintiff is the only reparation available in civil litigation. It is not long, however, before the potential size of an award becomes preeminent in the case. A maximum insurance coverage of $1 million is less compelling than if multiple defendants have three or four layers of insurance.

Plaintiffs' attorneys in smaller firms must decide how much of their company's money they can invest in the case or how much they can borrow to pay for the experts and research. Larger firms have more concise actuarial tools to judge the cost versus risk of litigating a complex case. In any event, attorneys tie the eventual size and depth of the expert witness and investigative pool directly to the potential revenues of a settlement or award.

The formula for estimating the value of the case depends on the depth of the pockets of the potential defendants and their insurance carriers. In my experience, attorneys look closely at the layers of insurance of defendant companies but, until now, have not adequately tied their ability to pay to their degree of culpability in the incident. That discussion is forthcoming. If the only defendant is a local auto repair shop with minimal E&O coverage, the plaintiff's story may be very brief and must stand on its own merits. If the defendant is a Fortune 100 manufacturer, the plaintiff's counsel will advise the client to retain the top-shelf experts as quickly as possible. The offshore manufacturer or service provider is also becoming a major factor in this cost/benefit equation. If the offending product's manufacturer is overseas and there is no corporate presence in the U.S., there is little appetite for litigating internationally or for pursuing vicarious liability. The international topic is discussed in more detail later in the book.

While compensation for the actual damages, reduction in future potential earnings, and the number of plaintiffs contribute to the potential value of a case, the deep pockets of the defendants and how spectacular the incident might be drive up the potential value of the case. This book adds another compelling factor to the potential compensation.

FBP investigations are audits of a defendant's design, manufacturing, and quality practices. Having performed such audits at more than 700 companies over 55 years, I have identified key indicators that give a very precise indicator of the "health" of an organization. The metrics include assessments of the effectiveness of vision, mission, values, design, marketing, manufacturing, customer service, quality, reliability, and cost of quality. The outcome of the assessments forms a proposal for Business

Process Improvement (BPI)[1] initiatives to help the company increase profits and/ or market share. These investigations lead to a list of deficiencies in the company's standard of care. Some reports are so scathing in their condemnation of the business practices of the defendant company that settlement quickly follows. The investigations often disclose critical deficiencies that were unknown to the senior management. The breadth of process nonconformities leading to the manufacturing of defective products is often staggering. I have documented inadvertent non-compliances with regulatory requirements that were undiscovered by the organization's quality department. I have also discovered an overt disregard for safety and quality that was part of a covert plan to increase profit and market share. These audits are conducted by examining the defendant's own documentation obtained via production requests.

The result of these non-traditional expert reports is that the defendants and their attorneys are blindsided with scientific data that is irrefutable evidence of negligence or inappropriate standard of care. Frequently, the fear of this information becoming public knowledge in court records drives the case to a rapid, confidential settlement. In one case, working for the plaintiff, the anticipated award became much higher in settlement after my testimony due to the looming threat of punitive damages in a jury trial. Conversely, the defendants were shielded from culpability in the case of the aerosol can (discussed previously), and the positive audit data helped their marketing strategy.

12.3 BUSINESS AS USUAL

Long before becoming an expert witness, I was aware that many of the companies that I audited for quality problems or business improvement potential were ticking time bombs of liability. Many organizations have undetected carcinogens metastasizing within their processes and procedures that manifest themselves only as inadequate profits or customer service issues. The pressures of day-to-day business very often mask subtle disasters in the making.

Manufacturers and service providers seldom look objectively at their processes, products, and services from the perspective of potential risk. It is not our nature to move 50,000 feet above our company and systematically analyze each step in our business operations, looking for potential opportunities for negligence. At best, most organizations grudgingly implement risk mitigation guidelines from their insurance carriers and consider these measures an imposition and added cost.

There are also a handful of companies that consider elevated risk a customary and acceptable cost of doing business. In my work as a quality control engineer and business consultant, I have witnessed company leaders who have checked their conscience at the door and built "acceptable risk" into their business model, as evident in the golf cart story. They build (or import) products that are of questionable quality and sold at discounted prices to an unsuspecting public, knowing that very few will complain and even fewer will attempt to sue the company. The inevitable lawsuits become just another cost on the balance sheet with no regard to individual or social consequences.

Even more egregious are the handful of business leaders who have contributed significantly to our recent financial crises. They methodically dismissed conscience and potential liability to create personal wealth and power. The eventual consequences of their actions may result in some form of organizational negligence and legal action, while the major consequence has the effect of destabilizing global economies and further creating distrust of corporate entities.

A profound example is the rise and fall of Enron. Enron Corporation was an American energy, commodities, and services company based in Houston, Texas.[2] Before its bankruptcy on December 2, 2001, Enron employed approximately 20,600 staff and was a major electricity, natural gas, communications, and pulp and paper company, with claimed revenues of nearly $101 billion during 2000. Fortune named Enron "America's Most Innovative Company" for six consecutive years. At the end of 2001, it was revealed that Enron's reported financial condition was sustained by an institutionalized, systematic, and creatively planned accounting fraud known as the Enron scandal. Enron has become synonymous with willful corporate fraud and corruption. The scandal also brought into question the accounting practices and activities of many corporations in the United States and was a factor in the enactment of the Sarbanes–Oxley Act of 2002.

To compound denial, most successful business leaders pride themselves on being problem "solvers." Few, if any, boast about being problem "avoiders." There are rarely, if ever, employee awards given for averting a major catastrophe that never happened. In fact, those who worked on the Y2K[3] computer problem were not given the proper respect because "nothing happened" on January 1, 2000, and their years of preventive work were dismissed as irrelevant and unnecessary. We do not learn enough about proactivity in business school. Our most robust current models typically work to minimize risk, not avoid it.

The current incarnation of "business as usual" includes the following rationalizations and half-truths:

- We cannot compete with foreign labor costs.
- We have little control over the quality and safety of the products we import.
- We must cut costs to be able to sell to the big box stores and retail chains.
- Our stockholders do not want to hear that business is down because of a recession.
- Labor unions make us non-competitive.
- It is not cost-effective to have consumers return defective products
- Technology changes so quickly, we must sell products and services that are not well-tested.
- Our products are safe because they are UL-listed.
- Creating a brand name in the market is more important than the product or service.
- We have an extensive customer service staff to deal with product and service issues.
- Our product is not defective. The consumer used it improperly.
- The Consumer Product Safety Commission is the gatekeeper for defective products.

- We have allocated a substantial reserve in our budget for warranty and recall costs.
- We are developing new products based on renewable resources that will solve the current issues.
- We cannot compete because of government regulations.
- It is impossible to hire and keep employees who are accountable for their work.
- Anyone can sue us for any reason (which abdicates responsibility).

Each of these topics contributes to defining products liability in the United States. If consumers tacitly abdicate accountability and business ethics, organizational negligence litigation will continue to thrive, and clever attorneys will find new approaches to assign blame. Unfortunately, they will also continue to find new ways to concoct frivolous litigation.

12.4   BLINDED BY THE LIGHT

Traditional forensic pathology is the science of determining the cause of death by examination of a cadaver. I have developed Forensic Business Pathology as the science of studying pathological diseases manifest in dead and dying organizations.

As a quality auditor and business consultant, I have been successfully examining the health of our clients' businesses to certify their fitness to be a supply-chain partner or to prepare them for strategic business process improvement initiatives. The case studies I have presented are rich in historical data on successes and non-successes in process and human improvement. Enlightened business leaders use the data points I provide to improve the processes that drive their key performance indicators continually. Over the last decade, I have added complementary data from another perspective: that of catastrophic business outcomes that resulted in products liability litigation.

As an expert witness in products liability and organizational negligence, I have amassed a host of compelling case studies of the most egregious business failures that lead to fatalities caused by faulty products and services. I now have undeniable proof that the warning signs of business diseases gone undetected can not only lead to financial calamities but can be the underlying cause of disasters that lead to human suffering and deaths.

My historical database provides irrefutable evidence that rework, warranty issues, customer service overhead, and productivity issues are symptomatic of undetected breakdowns in business processes and leadership. These symptoms are visible only to the enlightened observer as early evidence of congenital disease within their business. Without effective diagnostic tools, a carcinogen may have infected the body of their business with understated symptoms that are often unseen or ignored.

In the world of products liability litigation, forensic scientists work to develop compelling theories about the cause and origins of a disaster from the charred remains. Experts from both sides of a lawsuit will frequently disagree after examining the exact same evidence. In parallel, I study the business processes, quality standards, and available historical data to determine if the company that manufactured the product or delivered the service exhibited an appropriate standard of care to its customers. Our reports read like a business process or quality system audit. In this arena, however, I provide our quality audits to the most critical customers in the world: civil juries. The outcome of my reports and testimony is usually unassailable and decisive because it deals with measurable data instead of forensic hypotheses. I have mined this data and have made it the historical cornerstone of Forensic Business Pathology.

My unique tools are employed by very few business entrepreneurs, executives, and those with the vision to inoculate their businesses from potential liability while paying for the immunization by dramatically reducing rework, warranty costs, and the overhead of customer service. This lack of awareness and vision will keep the court system fertile with products liability and organizational negligence lawsuits for decades to come.

Whether a manufacturing or service company is a start-up or 100 years old, a mom-and-pop boutique, or a giant conglomerate, most businesses start with an idea and an entrepreneur. All entrepreneurs share the traits of being visionary, rebellious, undaunted, risk-taking, and internally driven. Successful entrepreneurs have been able to parlay their talents with other complementary resources and fateful timing to bring their vision to market. (There are no unsuccessful entrepreneurs, only those whose timing was inappropriate or whose resources ran out prematurely.)

In the start-up phase of a company, the entrepreneur surrounds himself with like-minded friends and supporters who work tirelessly to get the product or service off the ground. Most often, however, the road from concept to market is mined with explosive problems. They must make pragmatic decisions daily that inevitably erode the "ideal" version of the product into one that is marketable and cost-effective. It is all too often heard: "That will have to work for now. We can clean it up in the next revision." The reality is that there is never a scribe in the room to document the decision and its potential future impact. The entrepreneurial spirit has little bandwidth for reflection, and the compromise is seldom revisited. A list of seemingly unrelated events may build into unrelenting customer issues, including a time bomb of liability.

Why is this pathological entrepreneurial behavior so predictable? I have yet to meet an entrepreneur whose start-up strategies included documenting a shared vision, building an inner circle of trust, defining fitness for intended use, enacting irrevocable standards of customer care, establishing competency and training requirements, building a scalable business model, defining supply-chain management procedures, implementing formal customer satisfaction metrics and, most importantly, defining the personal accountability required for every individual in the organization. Some of these processes are painfully implemented after a crisis. Most are never addressed. In either case, the lack of one or more prevents most start-ups from achieving their greatest potential.

More critical to success and survival, the lack of ineffectiveness of any one of these tenets can be the catalyst for unwanted, unplanned, and unforeseen products liability. As businesses move from start-up to initial success, the operational processes most often evolve pragmatically and by tribal knowledge. That is, a small group of people work through issues and select seemingly viable approaches for moving forward. That same group works through daunting steps of problem-solving, achieving some pragmatic outcome for each stage of development.

The paradox is that the small group that worked through the steps of trial and error forgets the painful steps when it moves on to the next problem. They have brewed a palatable pot of soup but failed to document the exact ingredients and their proportions. Neither was the cooking time, temperature, nor attributes of ingredients codified in a recipe. It is erroneously assumed that the same chefs will always be together to cook the same pot of soup.

Clients typically marginalize my suggestion as an entrepreneur, team member, consultant, and quality analyst to capture the success steps for future replication because "we are already behind schedule." "The investors are breathing down our necks." "The customer just wants a product in his hands." The client would reject altogether an FMEA[4] event. The prototype product or service always becomes the model from which future products and services evolve, yet the evolutionary history of attributes, components, and process steps is lost in the haste to get to market. In products liability cases, a genetic or viral flaw in the original design or subsequent revisions violated the foundational tenets of the designers.

As companies grow, the original group of chefs often disbands, moves on, or changes roles and is replaced with employees who have no knowledge of the evolution of the product or service, the values of the designers, and the inbred developmental errors. There may be one tribal member in each area who attempts to hand down the customs and history, but inexorably, that knowledge is diluted and lost forever. Bottom line, the inevitable and self-destructive common thread is the failure to have a strategic plan, follow it and document the "as is" state as you evolve your entrepreneurship.

In too many cases, the bright light of financial success blinds otherwise enlightened leaders to the potential liability they are building into their products and services.

12.5 CAUSE AND EFFECT

In 1892, Milton Florsheim never anticipated that his shoe company would grow into an international force in men's footwear. He and his father were artisans, and they made the Florsheim brand synonymous with well-fitting shoes that would last for many years. They also created a marketing breakthrough in men's shoes when the company began opening exclusive retail outlets that sold only their shoes.

Eventually, Florsheim shoes became a retail commodity instead of an exclusive company brand. The original Florsheim Company no longer manufactured the brand. The new owners adopted their name (The Florsheim Group) and reputation, but the product had become a shoddy substitute for what was formerly a superior product.

Fast forward to a headline from business journals in 2002:[5]

Beleaguered shoemaker Florsheim Group Inc. threw in the towel Monday, seeking Chapter 11 bankruptcy protection after failing to make a $1.2 million payment to note holders last week. The move was hardly unexpected given Florsheim's mountain of debt and rising losses, retail experts say. However, a potential buyer for the company's assets was a surprise. Florsheim said it has agreed to sell its wholesale business and assets to a Milwaukee footwear firm that is run by descendants of company founder Milton Florsheim.

Milton never envisioned his descendants selling the business, with the new owners using his name and reputation and then sacrificing quality and workmanship for market share and profitability. The fact that his heirs bought it back is very atypical. The more common business model includes continual dilution of the original product quality in favor of making products faster and cheaper and selling them to box stores and warehouse outlets. After destroying the founder's reputation, the buyers moved on.

The common business model, which I alluded to earlier, works as follows: A start-up entrepreneurship evolves through hard work and dedication to producing a product or service of great value to the target consumers. The innovators learn that being the best and first to market are the hallmarks of featured articles in business journals. Esteem and reputation stimulate growth and market share. The business matures financially and in capacity until it plateaus at an arbitrary ceiling, reaching its highest level of success with the talent available.

At this three-to-five-year juncture, the founders are weary from the journey and are looking to either throttle back, enjoy the fruits of their labors, or hand over the administrative duties to others so they can return to their passions. There are several options that will play out at this point.

One of them is to sell the company, take the money and run. This may have been a strategic goal from the onset or one of opportunity. The new owners are buying a brand, reputation, product, service, existing market, innovative technology, and future growth potential according to a very precise financial success model. In their due diligence, the future owners have already fleshed out a plan to lower costs, increase profitability, and grow market share. I have observed some acquisitions where the buyer's business was merely buying and selling companies and had little concern for the products or services.[6] These groups follow a carefully designed success blueprint for systematically increasing revenues, profits, and marketability by riding the crest of the reputation wave of the seller while continually cheapening the product or service. Such was the case with the Florsheim model and many others.

A second avenue is to hire managers to run the business so that the inventors can get back into R&D and out of finance and marketing. The personnel engaged in taking over the day-to-day operations are seldom selected for their ability to codify and replicate the success model they are employed to administer. Instead, prior business experience and success form the basis for hiring. Unfortunately, how they achieved prior successes is seldom obvious until they are on the ground and are making changes to the previous business model. They map a growth strategy based on their own experiences and vision, not on those of the entrepreneur who retained them. By this time, the founders have abdicated their responsibility for running the business and returned to R&D while the ship is being steered on an unknown course into uncertain waters.

A third scenario adds apathy and abdication. The company is making money. The founders have accumulated some sustaining level of wealth. The hired-gun CEO, CFO, and marketing manager are busy and continually reporting the addition of new customers, market segments, and increased profitability. Covertly, the new bosses redirect the manager of sustaining engineering to start cutting costs from the product by substituting less expensive components. The quality manager is encouraged to relax their standards to make the product more competitive. The entrepreneurs are busy developing revision 10.1.5 and do not see the slow and methodical decay of their product or service until they are blindsided by a customer complaint of epic proportions. By then, a tumor is growing in the belly of the beast that has created malignancies in the vital organs of the company. In the blink of an eye, the company finds itself in the throes of triage and damage control over an unseen defect that has allegedly caused great harm to a customer.

If the problem is severe and pervasive enough, it could be the end of the entire enterprise. The effect can be a financial disaster and individual liability, while the cause may be no more complex than apathy or abdication of the founding principles of the company. Business success is often the cause of products liability and organizational negligence. Being unaware of the actions of entrepreneurs can lead to an early demise.

12.6 PROCESS CONTROL

Experts in quality management and organizational excellence have studied the complex and independent systems that form manufacturing and service businesses. These practitioners focus their efforts on performance, safety, and reliability improvement, leading to increasingly more robust products and services. The impetus for enlightened business leaders to invest in these programs typically lies in continually raising the measures of their success. To conduct detailed system analysis, experts dissect the individual processes for detailed examination and then reconstruct them into a process road map. Defining process control in the context of products liability, I will change focus from process Improvement to process assessment and detection of processes that run the spectrum from being sub-optimized to completely out of control.

Proc·ess (noun). A systematic series of actions directed to some end: to devise a process for homogenizing milk.

Referring to Figure 12.1, the input of a process is usually a combination of specifications and constituent parts. The activity is the value-added steps of the process. The output is a measurable object or service that is a component of a larger system. Once it has been verified as compliant with the requirements of the process, it is ready to move on to the next process step. An example might be the process of shaving. The input is a stubbly face, shave cream, a razor, and a lifetime of training and experience. The activity is the physical process of putting on cream, shaving, and cleaning up. The output is a smooth face. Measurement of results is an examination in the mirror for quality of workmanship and appearance of the face. The next operation might be getting dressed.

Moving from the individual process to the larger system, enlightened businesspeople construct a process model that is a closed loop of ensuring the effectiveness of a process. Continuing the previous example, brushing our teeth might be a component of getting up and readying ourselves for the work process. Going to work might be a subcomponent of the overall work process. Our careers are then constituent parts of the process of life. I then construct the process model from the atomic level to the galactic level, with no one component being any more or less critical to the success of the entire system.

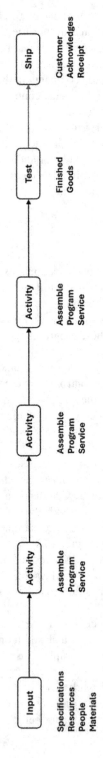

FIGURE 12.1 The Conventional Process Model

This figure depicts the flow of a process from left to right and from start to finish.

As discussed earlier, for processes to be repeatedly successful, the process owners must work diligently to minimize process variability. Let us put the concept of process control and process variability into a practical example.

Think about how many times you have told your children to mow the grass. Were the results you obtained exactly what you expected? If you do not have children, think about the same scenario with your parents telling you to mow the grass. How many times were there missed spots, clippings not bagged, the mower left in the driveway, or the edging not done? Who trained you and your children to mow the grass? Hasn't it mostly always been by assumed observation and some intuitive sense of what a properly manicured lawn looked like? What would mowing the grass look like in a structured process model?

12.7   MOWING THE GRASS

Vision: To retain the aesthetic beauty and intrinsic value of our home.

Mission: To maintain a consistent appearance in the vegetation surrounding the house and to ensure a safe home environment, free from pest and rodent infestation.

1. Put on heavy boots and safety glasses.
2. Inspect the mower. Verify the air filter is clean, and the blade is tight.
3. Be sure the level adjustment is set properly and all wheels are at the same height.
4. Check the engine oil. Add, as necessary.
5. Fill the gas tank. Get gas if there is none in the gas can.
6. Start the mower and mow in a logical sequence.
7. Be sure to cut all areas that you can reach with the mower.
8. Inspect the edger. Add gasoline/oil mixture. Check to be sure there is enough cutting line in the edger and that it is properly wound.
9. Run the edger around all perimeter areas. Be careful not to damage flowers, trees, or decorative trim.
10. Rake up excessive grass clippings. Place clippings in the compost bin.
11. Inspect your work and verify the job was done per item 1 through 10. Report any damage that occurred during mowing and edging.
12. Clean the edger and mower and return them and all tools to their storage spaces. Report any observed problems with the mower or edger.
13. Keep track of when the oil needs changing in the mower or if there is a sign of blade wear.
14. Report any observed problems in the flowerbeds, with the trees and shrubbery, or with insect or rodent infestations.
15. Mark your calendar for the next scheduled mowing event.

While most would scoff at the thought of codifying a process for mowing the lawn, training in the procedure would cause the process to be consistently performed correctly. Would this process not run smoothly without continual intervention from management? Would not such a process prevent lawn disease and termite and rat infestations? Would not a methodically groomed yard increase the value of the home at the time of sale?

To reiterate, while minimization of process variability is a given fundamental tenet in manufacturing, the comprehensive collection and study of the data needed for effective diagnoses is seldom given the priority required for effective liability avoidance. Enlightened business leaders are rarely aware that they are building in liability risks until processes drift so far out of control that they cause financial or human catastrophes.

The data used to maximize productivity is directly applicable to failure investigations. The same data from process analysis and human performance from the perspective of potential failures and anomalous behavior becomes an evidence trail that explains scientifically how an organization permitted a defective product to reach customers. In products liability litigation, such studies reconstruct the actual manufacturing environment. It is possible to recreate the series of events leading to the manufacture of a particular product with similar precision as forensic scientists do from a disaster scene and trace evidence. Instead of using scientific instruments, a business forensics expert uses advanced proactive auditing techniques to examine process procedures, production drawings, workmanship standards, training records, test results, and quality records. This expert compares them to the depositions or testimony for consistency and efficacy. The entire process tells a compelling story for the plaintiff or defense.

Did the patient who died of a massive infection succumb because he did not brush his teeth correctly or because of the negligence of others? While the plaintiffs need an expert to determine the exact cause of death, they need a Forensic Business Pathologist to interview the family about the deceased's oral hygiene process to quickly construct a scientific process map of what the likely events were that led to the death of the patient and where the culpability truly lies. Attorneys can save thousands of dollars spent on opposing experts opining that the dentist did or did not exercise an appropriate standard of care and the case could be dismissed without merit.

Keep in mind that this tooth brushing example is simply an analogy. In the business world, Forensic Business Pathology can be an invaluable tool to ensure your processes are synchronized, efficient, and cost-effective, while avoiding risks that can lead to litigation.

12.8   STANDARD OF CARE

"Standard of care" is used many times in this book. However, there are no legal lessons on how attorneys use this term in litigation strategy. It is my intent to put into context how the contents of this book are relevant to proving an appropriate or inappropriate standard of care and what impact that evidence can have on the outcome of a lawsuit. Provided here is my own understanding of the standard of care and how I have successfully applied Forensic Business Pathology to case after case to reach more appropriate outcomes.

Standard of care is determined by the standard that would be exercised by the prudent manufacturer of a product, service provider, or the prudent professional in that line of work. At this point, the definitions become circular and convoluted because a judge or jury must decide if there was a specific *duty* of care requiring that the defendants adhere to a standard of reasonable care while performing any acts that could foreseeably harm others. Compounding that obtuse logic, I add the test of the "reasonable person," which requires establishing a standard against which any individual's conduct can be measured. Litigators use the reasonable person standard test to determine if a breach of the standard of care has occurred, provided a duty of care is provable. Product liability cases can drag on for years, just arguing over the meaning of manage and lead.

To bring order to this morass, let's sort this process into Boolean and conditional logic by beginning with the term duty of care. There is a universe of existing case law about the duty of care for most services and common manufactured products. While each side will argue which case law (that supports its position) is applicable and present its take on what determines the actions of a reasonable person, by placing the arguments into rows and columns, I can bring great clarity to the most convoluted failure scenarios.

To begin at a macroscopic level, I will attempt to define better the duty of care of those who sell products and provide services to consumers. I will propose that a consumer product is one that fulfills one or more consumer needs reliably and safely. The product must meet its stated and implied advertising/specifications and be fit for its intended use. For example, a consumer would purchase a toaster to brown a variety of bread products to the user's desired level of darkness. Its specifications are minimal. After centuries of toasters being on the market, a reasonable person would innately know what the product was for and how to use it. Aside from the obvious use with sliced bread, the size and depth of the chamber in which the raw product fits limits its capacity. A reasonable person would expect the toaster to work reliably for an extended time, based on frequency of use. The consumer would expect the product to be safe in the usual context of its use. I propose that the duty of care of the manufacturer is to provide a product that will safely and repeatedly heat breads and pastries to a desired level of doneness. Further, the duty of care of a toaster manufacturer is to provide a product that will not burn, asphyxiate, or electrocute a reasonable person when used for its stated and implied intent.

Yes, a first-year law clerk could make compelling arguments for the pros and cons on each item in the preceding paragraph. Please allow some latitude as I connect the dots into a coherent process map.

Continuing our hypothetical scenario, what standard of care should be applied to the toaster manufacturer based on its standard of duty? To answer that question, I will now begin to construct a compliance matrix to present a visual to enhance the words I will use.

I transform the ambiguities of duty of care and standard of care into a system of interrelated and interdependent processes rather than individual nebulous terms such as safety or reliability. The examination is as complete as a quality control engineer would need to perform tests and analyses that would validate the integrity of the subject toaster. Duty of care lists every reasonable standard that a consumer or safety testing agency would levy on a toaster sold for domestic use. Standard of care establishes unambiguous parameters for measuring duty of care. Requirement establishes the evidence by which standard of care is physically manifest. Measurement provides scientific tests and evidence trails that are facts and not opinions.

If the toaster manufacturer hired an expert to analyze perceived quality and customer service issues with its toasters, their report would be one of the work output products. Since most business executives have never dealt with the term's duty of care and standard of care, the first column would be labeled "Design Goals" and the second column "Objectives." Once the company agreed that this matrix contained every criterion that it needed to optimize and measure, a specific process improvement plan would be constructed and implemented.

In Forensic Business Pathology, this exact same matrix would reference specific evidence, examples, and parameters of compliance or noncompliance with the defined standard of care. It will be an exhibit in an expert report. Testimony and opinions would address each cell in the matrix. Instead of subjective opinions, an FBP investigation would conclude as follows (in reference to "meet stated requirements in the matrix"):

Compliant Outcome—This expert examined three exemplar toasters in the context of an informed consumer to determine if the controls performed as stated in the user's manual. Using six slices of Kroger's Best White Bread, the thermostat knob setting browned the bread very similar to the graphic on the toaster, from barely warm to nearly char. The switch that lowered the bread into the toaster and engaged the heating elements worked each time it was depressed, and the button specified for use in manually stopping the process also worked as specified.

Noncompliant Outcome—This expert examined three exemplar toasters in the context of an informed consumer to determine if the controls performed as stated in the user's manual. Using six slices of Kroger's Best White Bread, the thermostat knob had very little relationship to the graphic on the toaster. The test slices varied in warmth and color from room temperature to smoke pouring from the toaster when it was on its darkest setting. The switch that lowered the bread into the toaster and engaged the heating elements worked each time it was depressed, but the button specified for use in manually stopping the process had to be pushed several times to stop the toasting process.

As the expert report steps through each item and parameter of the matrix, an audit trail emerges that constructs irrefutable evidence of an acceptable or unacceptable standard of care. While the expert would expect to be cross-examined on the criteria used to judge the darkness of the bread, the body of evidence in the matrix, when analyzed collectively, provides a compelling story validating that the product was robust and fit for its intended use or there were unmistakable markers that the product was ripe for failure in normal use.

Using conditional statements in testimony prevents the subjectivity of the color of toasted white bread from becoming consequential. To wit: "The fact that the control knob allowed the bread to reach a temperature high enough to cause visible smoke and burning not only could lead to ignition of the surroundings of the toaster, but it could cause asphyxiation and burn the user when they tried to extract the burning toast." Dissecting that statement: If the control knob worked correctly, then the bread would not burn, and the user would not be hurt, or else they would be injured by burned bread.

Looking at the matrix as a holistic system, the color of the bread is inconsequential to the overall outcome of a compliant product, and it can become one fact that builds an irrefutable argument for an unacceptable standard of care in a noncompliant product. The processes defined herein remove the subjectivity from proving an appropriate standard of care, which is vital to reaching a rapid and cost-effective outcome for the case. More importantly, regarding the outcome of a case for the plaintiff, I have shifted the focus from a single failure that caused a single accident to establishing a pattern of neglectful behavior as usual and customary for the defendant company. For the defendant, this investigation can establish that the company has a history of being extremely diligent in its standard of care, so even a single event, in this case, was unlikely to be the result of actions taken by the company. Even in a very emotional case, facts should trump opinions because a long-term pattern of behavior can predict accidents in the making. The better news is that a trained expert can readily mine the evidence.

12.9   THE FOUNDATION OF FORENSIC BUSINESS PATHOLOGY®

At this juncture, placing the evolution of FBP into a historical context is critical to affecting a paradigm shift that will benefit the legal (and commercial) community in understanding business pathology. I was fortunate to be among the space pioneers at

FIGURE 12.2   The Author at Mission Control Houston circa 1972

The author was photographed at the console in one of the support rooms at NASA's mission control center in Houston.

NASA's Mission Control Center from 1966 to 1980, where I learned the foundational tenets of what would become FBP.

NOTES

1 One of my former colleagues became the creator of the Dell International Business Process Improvement initiative. Twenty-five years later, it is still active as dellrefurbished.com.
2 From Wikipedia.
3 There was a theory that vast numbers of computerized devices would stop working when year 2000 came.
4 Failure Mode Effect Analysis.
5 Available at: https://www.company-histories.com/Florsheim-Shoe-Group-Inc-Company-History.html
6 This was evidenced in the defective outlet case.

13 Flying the Mission

13.1 MY LIFE IN MISSION CONTROL

What turned out to be the understatement of April 1970 was a wake-up call for those of us working on the flight of Apollo 13. Until the imperative "Houston, we've had a problem" disrupted the customary air-to-ground chatter, it had been on a nominal flight plan for what the press had dubbed just another routine lunar space mission. Three astronauts were two days into a voyage from Earth to the moon when disaster struck.

I was having a cup of coffee, reading the mission log from the day shift, and listening to the air-to-ground loop on the communications console. Those first words from astronaut Jack Sweigert caused me to sit up smartly in the chair, set aside the shift log, and put a fresh light on my pipe.

I turned the volume up in time to hear the ground controller reply, "Say again, please." Mission Commander Jim Lovell repeated the declaration, "Houston, it looks like we've had a problem." The astronauts reported they were rapidly losing power in Odyssey, the command module. There was also visual evidence that they were venting gases into space. Either problem was potentially life-threatening. Together, they were anomalies beyond those ever envisioned or simulated. I would learn later that a required maintenance procedure performed by astronaut Jack Sweigert had precipitated an explosion in the service module of the spacecraft.

I immediately selected several more communications loops to monitor the dialogue between the various mission specialists. I leaned into the speaker, hoping to follow the cacophony of voices and make some sense of what had happened. The plume of smoke from my pipe must have looked like the stack of a steam locomotive, keeping time with my elevated pulse.

The Quality Assurance (QA) Office was directly across the hall from the Second Floor Mission Operations Control Room (MOCR), Mission Operations Wing (MOW) in Building 30 of the Johnson Space Center. That MOCR was Mission Control for Apollo 13 (there was a duplicate control room on the third floor that was active in case of a systems failure on the second floor). I was a 25-year-old Quality Control Engineer assigned to the swing shift for this mission. There were six of us in the QA department of Philco, the prime contractor for Building 30. We staffed the QA office 24/7 during missions.

When no space missions were flying, my regular duties included inspecting the ground support hardware for defects, monitoring qualification tests, and performing capability audits of potential suppliers to NASA. During a mission, my responsibilities also included being part of the Technical Operations (tech-ops) team. Our assignment was to be available in case there was a malfunction in any of the ground support hardware in Building 30. Should a problem occur, we were responsible for monitoring repairs, witnessing testing of the repairs, and re-sealing the hardware

DOI: 10.1201/9781003535621-18

with tamper-proof devices. With the rigorous processes we used to build, inspect, and test the hardware before missions, there was seldom a need for us to do our jobs during a flight. Since the Johnson Space Center began controlling manned space flight, QA's greatest contribution to lunar missions had been keeping the shift log current and staying close to a communications console in case we were needed. This was only my second opportunity to support a live mission, so I was more involved and loved every minute of it.

I became aware of a flurry of activity in the hallways outside the MOCR but was engrossed in the communications chatter. The mission specialists were close to a diagnosis of the explosion and what the impact of the problem would be on safety and mission success. I followed the developing scenario of an oxygen tank exploding and causing the power loss in Odyssey. The Apollo spacecrafts used cryogenics both for breathing and creating electricity. This double whammy had dire consequences.

Although never done or simulated, mission specialists shut down the command module to conserve the remaining power for reentry and use the Lunar Excursion Module (LEM) as a lifeboat. This power-down procedure was unprecedented. While they devised a rescue plan, the astronauts moved into the LEM as a makeshift life-boat. By now, I had silenced the speaker and put on a headset for greater clarity. I was chewing a hole in my pipe stem and preparing myself emotionally to be part of the recovery team. Although I had no specific duties, I was ready to respond to a call to action and join the beehive of activity that was taking place on the communications loops and in the hallway.

All sense of time had vanished. Fortunately, the hours of simulation I endured as training for problems snapped me into reality. A familiar voice overrode the chatter on the communications console. "QA—Tech-ops, my loop." Tech-ops were the only functioning two-way circuit QA had, and the volume was significantly above the drone of the other voices. Tech-ops coordinated all the maintenance and configuration work on the ground support hardware and directed repair efforts during missions. "Tech-ops—QA—Go" was my trained reply. "Who is on duty?" the voice asked. "Taormina" was my only retort. In a rapid succession of two-way dialogue, it was determined that the only Polaroid camera in the building was in our office. "Get down to Sub-stores and bring all the film back to your office and stand by," was the final transmission before I relinquished my headset and laid my pipe down.

In the opposite corner of the second floor was a storeroom of operational supplies to support the needs of Mission Control. As I exited my office, for the first time, I became aware that the activity I had heard was the sound of programmers and mathematicians setting up schoolroom chairs in the hallway. There were at least a dozen of them, by this time, thumbing through computer printouts and manipulating slide rules. I would learn later that they were performing calculations to determine the best option for returning the spacecraft and the three astronauts safely to Earth. By the end of the night, the rectangular hallway was crammed with a staff of specialists concocting various rescue procedures in real time.

There must have been 20 rolls of film on the shelves in Sub-stores. I filled out the appropriate requisition form and hoofed it back to the office, film in hand. Almost immediately, a familiar face breached the doorway and asked, "Are you the QA guy?" As it turns out, he looked familiar because it was senior astronaut Frank Borman.

Although I worked a few feet away from the MOCR, I almost never interacted with astronauts or flight controllers. Meeting Colonel Borman was a real treat for me. "Yes sir, that's me," was my only reply. He told me to get the film and the camera and follow him down the hall to the makeshift tactical control center. A crowd was gathering in a staff support room that was normally vacant during a mission.

Whether you recall the televised events of those few days or you have seen it portrayed in Ron Howard's compelling (and highly accurate) movie, the rescue plan was hatched by astronauts, flight controllers, and mission specialists in a smoke-filled room on the second floor of Building 30. In mere hours, they had to contain the disaster and create a viable rescue plan. In parallel efforts, they would solve the issues of keeping the astronauts alive, returning them to Earth, and making a successful reentry and landing in a spacecraft with little power and even less air to breathe.

One of the accurate details in the movie Apollo 13 was the chaotic moment in the tactical control center when the lamp in the overhead projector burned out. Rather than hunting down a new bulb, they reverted to chalk and a blackboard to brainstorm the problems. My job was to take Polaroid pictures of the calculations and diagrams they were drawing on the board. After they were through with a particular scenario, I would wash the board, and they would start over again. I turned over the photos to the appropriate specialists to convert the data into a revised mission plan.

I was the designated photographer for the rest of my shift. When I left the tactical control room, I noted my activities in the shift log, collected my pipe, and prepared to hand over the special access badge to the next QA representative. The passing of the restricted-access badge (that gained us access to every corner of Building 30) was a ceremonial indication of a successful handover, mimicking the process the various teams of flight controllers conducted at the end of their shifts.

13.2   SPLASHDOWN

After the successful landing and recovery of the Apollo 13 crew, I had our usual series of splashdown parties and debriefings and began reconfiguring the third-floor MOCR for Apollo 14. Within Mission Control, we relived our respective roles in the recovery effort over coffee, but there was little sense of the impact those few days had on our lives, our careers, and our national history. In any event, life returned to normal, and I was soon back on the road performing capability audits of potential new suppliers to NASA.

Age and wisdom have caused me to be more appreciative of the pioneering work done in the 1960s and 1970s. Being young and idealistic can be synonymous with being oblivious to historical significance and potential danger. Youth also includes a certain naiveté that masks the overwhelming odds against accomplishing impossible tasks, such as the rescue of Apollo 13.

For many years, I was unaware of the personal and professional impact of my involvement in, and limited contribution to, that moment in history. Even though I was often asked questions about what it was like to be in Mission Control during Apollo 13, they were answered with technical data, statistics, and process details. There may have been some ego woven into the answers (and, certainly, the stories have become embellished over time), but we just did what we were trained to do.

After the adrenaline rush of the first hour of monitoring air-to-ground and flight controller conversations, the realities of my training became obvious when I went to Sub-stores and dutifully filled out a requisition form for Polaroid film. Although the problems were immediate and the situation was grave, I just did what we were trained to do. Why else would I have taken the time to fill out a requisition form amid a crisis? **The answer is that effective process procedures and training prevent anarchy and wasted effort.**

We found tactical solutions to the problems at hand and implemented them in the same way we found a strategic solution to the challenge from President John F. Kennedy that "this nation should commit itself to achieving the goal, before this decade is out, of landing a man on the moon and returning him safely to the earth." For centuries, man has dreamed of traveling to the moon. We just made our minds up and did it.

13.3   LESSONS LEARNED

In the ensuing ten years after Apollo 13, I was fortunate to have been assigned to perform capability audits of organizations such as Digital Equipment, Data General, Ampex, Hewlett-Packard, and Sangamo. I also conducted audits at hastily formed companies that intended to take advantage of NASA's policy of giving preferential treatment to minority business enterprises. Many of the latter were viable businesses, but some were obvious frauds perpetrated to exploit minority businesspeople and unfairly win government contracts.

As prescribed by our contract with NASA, I used the same quality audit checklists for all companies. The quality requirements and checklist contents were in the public domain, and all the companies had prepared for my visit and choreographed a tour of documents and procedures as evidence of compliance. Each exhibit provided would correlate to a check box on my forms. At the end of a typical audit, most of the requirements had been marked as compliant, and approval was automatic.

Constantly influenced by Apollo 13, however, auditing suppliers to the checklist began to make me uneasy. It was the same disconcerting feeling as when a salesperson describes their fabulous warranty and extensive service department. Why do they need a fabulous warranty and extensive service department if the product is inherently reliable? This discomfort was caused by my realization that a compliant checklist at Hewlett-Packard and a compliant checklist at Bubba's Electronics Company were significantly different. It was apparent in touring HP that they had a quality manual and personnel following written procedures. So, it was at Bubba's. The subtle difference was that at HP, the people were obviously following the documentation as trained. At Bubba's, the records were completed to indicate compliance, but the workers were operating by tribal knowledge, not written procedures. What they claimed to be doing and what I observed them doing were often different. In some audits, I was prohibited from talking to the workers, only to the quality manager or salesperson. These conflicted environments always caused my caution and warning indicator to glow red, but I had little choice but to follow the checklist.

I had spent the previous week auditing companies that were proposing to fabricate custom-printed circuit boards for us. Fabricating a circuit board requires chemically etching copper away, with powerful acids, from a "sandwich" of copper with laminated fiberglass in the middle. It also involves plating metals such as nickel, tin-lead, and gold on exposed copper. The entire process is very nasty. All the companies I visited had met the requirements of the checklist.

A Kairos Moment[1] made me aware of the difference between compliance with a checklist and meeting the needs of the customer. Reviewing the checklists from the previous week evoked a moment of great clarity. I had two identical checklists in front of me, yet the companies they represented were vastly different.

Let's call the two companies I visited Alpha and Bravo. Alpha's facility was typical of most circuit board shops. The air was thick with the choking smell of acid. The wooden floors were wet with spillage. The employees wore dirty smocks with holes burned through by the acid. The quality manager at Alpha quizzed me about how closely I inspected the tolerances for under-etching (leaving too much copper), over-etching (removing too much copper), and plating thickness (the amount of nickel, tin-lead, or gold added). This line of questioning made me wary that they did not always produce a product that met our specifications. He hurried me past barrels heaped with scrap boards and a locked cage with discrepant products awaiting approval of quality concessions by their customers.

Bravo's plant had no objectionable odors in the air. The plating and etching tanks were in containments that minimized spillage. The employees wore clean blue smocks, safety glasses, and gloves. There was no evidence of deterioration in any of their clothing. Bravo's quality manager asked only technical questions and if I preferred the circuit boards individually wrapped or in stacks of ten. His questions left me with a sense that they wanted to meet our stated and implied needs. There was a noticeable absence of barrels of scrap material or a withhold cage for discrepant products.

Under the microscopes in the two respective QA labs, the etching and plating of the sample circuit boards offered for my inspection appeared to be identical. In the Alpha QA lab, an old and rattling window air conditioner made it difficult for us to carry on a conversation. I did not sit down because of the discoloration on the chair seats. The brightly lit and environmentally controlled QA lab at Bravo ensured a clean and stable environment for conducting precision measurements. The quality manager at Bravo informed me the controls were necessary because the fiberglass substrate can absorb moisture, causing inconsistencies in measurements. The furniture was older, but there was no hesitancy about taking a seat.

Bravo and Alpha had identical checklists, but Bravo was clearly a superior supplier to Alpha. There is a significant difference between providing a minimally compliant product or service and providing one that consistently delivers value to the customer.

I was confident that I could receive acceptable products from Alpha, but the evaluation of their processes also made me anxious that we would receive just enough discrepant products, discovered at our receiving inspection lab, for us to deal regularly with rejected materials and tightened inspection controls. When suppliers have a history of delivering nonconforming commodities, there is a greater possibility that a random defective product might reach the user and lead to a latent failure.

My expectations from Bravo would be shorter inspection times when receiving because of a consistent history of zero defective parts. It would also defy all rules of quality assurance to expect a defective part to find its way to the end user. Which was the better value for the same price? In the case of the circuit board suppliers, which provided NASA with the lowest overall cost of ownership and the least chance of latent failure?

My narrative report explained my observations and conclusions and gave a "grade" to potential suppliers Alpha and Bravo. The narrative proved to be a valuable addition and was forwarded to the purchasing department, which placed Bravo on the interim approval list and set aside Alpha. Alpha called weeks later to see if they were going to get a sample order. The purchasing manager explained that they did not meet our quality standards and would not be receiving any sample orders.

Alpha, obviously perturbed, called our NASA contract monitor and complained. He knew it met the checklist but was not on the approved supplier list. How could that be? Our contract monitor did not have an answer. That precipitated one change in policy: NASA would send an engineer along on the audit trips for a few months and observe the narrative and grading system. Within six months, NASA approved the narrative and an objective grading system for ground support hardware suppliers. Alpha never made the cut as an approved supplier.

13.4   A KAIROS MOMENT

Following that victory, the leaders wanted me to explain to the world what criteria I should publish to objectively evaluate the total cost of ownership of products and services. In other words, what indicators could I identify that would objectively point out the perils of Alpha and Bubba and the merits of HP and Bravo? What attributes observed at HP and Bravo would predict that they would be a reliable and cost-effective supplier?

That challenge was like trying to explain something subjective, such as why pizza tastes better than celery. Some serious introspection led me back to the roots of the Kairos Moment, Apollo 13. How did we just do what we were trained to do and rescue three astronauts from certain death as routinely and flawlessly as a Swiss watchmaker building his thousandth precision timepiece? What were the lessons of Apollo 13 that gave me the clarity to establish pragmatic rules for evaluating the effectiveness of a company's ability to produce acceptable products and services?

Lessons from disaster can illuminate a clear path to follow to avoid future disasters. There were specific lessons learned from Project Apollo that, by design or by illumination, caused a business model to evolve, making excellence routine and failure not an option. Apollo 13 dramatically validated our methodologies by demonstrating that trained and motivated individuals can perform heroic and impossible feats when confronted by certain disasters.

A word processor would have sped up the process, but I spent months trying to correlate the lessons of Apollo 13 to auditing methods. There were reams of handwritten and typed epistles on establishing and evaluating business processes, many erased and redrawn lines, and boxes on diagrams of process flows.

The operational procedures and standards at Philco dealt with establishing command and control over employees and processes to ensure successful outcomes. The manuals were thick with inspection and detection, workers and inspectors, crime, and punishment. There was no documented evidence to support any observations of what made an organization truly successful. Our tongue-in-cheek rule of thumb had been that when the weight of the paperwork equaled the weight of the electronics hardware, we were ready to fly a mission. Surely in the mass of reference materials, there would be the enlightened teachings that supported humankind's greatest technological feat, traveling to the moon and returning safely to Earth.

My research led only to more quality control tools and detailed product and process specifications. There were no epiphanies to be experienced in these libraries. The data was contained in the events leading to my discovery of the pragmatic methods I had adapted for supplier evaluations. I began mapping my own history and how it had precipitated the enhanced business passion and curiosity within me.

This led me directly to the root of the driving force, the famous speech by President John F. Kennedy that included the challenge that "this nation should commit itself to achieving the goal, before this decade is out, of landing a man on the moon and returning him safely to the earth." The President was obviously talking specifically to me in this address. He was issuing a challenge to get my engineering degree and be part of making real the vision he dreamed of for us as a nation. I started college in the fall of 1963, so I would not be in the industry until 1968, so I only shared the vision in the early days of Project Apollo. The only logical solution was to leave the family home in Queens, New York, and move to Houston. While I went to college, I would be able to work at NASA and be part of the team as early as 1965.

This brought me to the Cullen College of Engineering at the University of Houston, finding employment at Philco a year later. My enthusiasm, not credentials, got me an entry-level electronics technician position. Our group assembled, wired, and tested the consoles and ground support hardware for Mission Control and most of Building 30.

I was contributing to President Kennedy's vision. His clear, concise, and resonant call to action was at the center of every activity I performed in my 14 at NASA. In fact, we had very few "employees" at Philco. There were only a few working there just to collect a paycheck. Most had a passion for space exploration. Quality control inspections were self-imposed because workers were so highly motivated toward mission success that making a poor solder connection or not checking the accuracy of the wiring was unthinkable to our mission-driven workforce.

The word "mission" kept coming back. The volumes of quality control requirements in our library had almost nothing to do with the precision workmanship we routinely achieved. Our reliable work output was a direct result of the mission statement provided to us by President Kennedy. Our engineers designed it for 99.95 % reliability. Statistically, that meant NASA would expect to have five failures in every 1,000 missions for pioneering technology, which is an acceptable risk factor. JFK's mission statement was integral to the engineers' tireless hours of designing for mission success. We also employed double and triple redundancies when flight safety was at risk. Along with the design engineers, the technicians, installers, and support personnel also shared the mission statement when conducting their daily jobs. The closest function we had to a warranty department was the quality engineers staffing their posts during a mission. I seldom had anything to do, making us prototypes for the lonely Maytag Repairman.

Finally, the lessons of Apollo 13 were becoming obvious. A clear, concise, and compelling vision and mission statement was the foundation for producing highly reliable products. All the rules and policing in the world could not accomplish what self-motivated individuals working toward a single common goal were capable of doing. Further, these same people, operating in crisis mode, could accomplish impossible feats in record times. Motivational training or monetary incentives are meaningless to self-directed individuals who have learned from a Kairos moment.

During Apollo 13, an imperative from head flight controller Gene Kranz embellished JFK's mission statement. His stated directive to the Apollo 13 team was that "failure is not an option."[2] By his leadership and presence, he inextricably etched these prime directives into every fiber of our beings and drove every action. Our dedication to delivering a compliant product on time, within schedule, and with no defects was not a contrivance of some clever management scheme. It was intrinsic, genuine, unassailable, and not for sale at any price. During the days immediately following the explosion aboard Apollo 13, many key individuals did not sleep, did not receive any extra pay, and would not have gone home a minute before their portion of the rescue was complete. A compelling mission statement can be this powerful and even awesome in the right environment.

Lesson #1 from this Kairos moment is that a clear, concise, and compelling mission statement, championed by a true leader and owned by everyone in the organization, is the cornerstone of business success. Organizations that have mission statements, value statements, and quality policies developed by the marketing department and appear only on the wall in the lobby and in the annual report are missing one of the most powerful success tools available. Business leaders cannot develop a vision as powerful as JFK's, but Gene Kranz was a senior manager who did not mince words. He made "failure is not an option" a driving force that we all could live without coercion or monetary incentives.

Once I laid the cornerstone, the rest of the lessons began to fall into place like the last few pieces of a jigsaw puzzle. The motivation of the Apollo 13 rescue team made it obvious that the successful rescue was a direct result of training and following process procedures. After identifying the result of the disaster, we immediately looked at our documented procedures, flight plans, reference documents, and checklists for guidance. Rather than abandoning our training, we followed the sequence of the documented procedures and modified them to deal with the new scenarios. We did not look for blame. We did not file a suit against the manufacturer of the defective oxygen tank. We did not waste time looking for the root cause of the problem. We would analyze the remedial options after the crew was safely home. We did not panic, circumvent safety, or try to launch a rescue mission that might have killed even more people.

After the initial shock of the disaster, cool heads prevailed and looked to successful procedures for guidance under the new circumstances. It was like various training videos on how to escape from a submerged vehicle in a lake or river. The trapped person must stay calm, allowing pressure to equalize before opening a window and swimming out. Alternatively, that person could panic and try in vain to open the door against the pressure of the incoming water. As police and emergency service personnel know, the only way to survive is to follow established procedures in times of disaster. Panic or reactive behavior will only make the disaster more destructive.

How does Lesson #2 translate for other organizations that are not flying to the moon? It clearly makes the case that companies must have a mission or mandate for success from leadership. They must have documented process procedures in place that give guidance for their daily operations. Training must be at a level that makes following the documented procedures key to accomplishing the mission statement. The procedures must be alive: they must always reflect current and changing conditions because of continual improvement and changing requirements. Continually training the workers in their content and importance is essential to long-term effectiveness. Airline personnel are continually in training and being drilled in simulators, even though they fly several days a week. Their written procedures are so robust and effective that the crews cannot lapse to reactive behavior, especially in a time of crisis. Experts have documented every potential anomaly in passenger airline travel, simulated and trained repeatedly. As a result, the airline industry enjoys the best safety record per passenger mile traveled.

Lesson #3 is that no mission statement or documented system of process procedures can be effective unless the environment exists to encourage entrepreneurship, even though entrepreneurship might appear to be paradoxical, especially in an environment such as NASA.

An entrepreneurial environment is one of self-motivated individuals working toward a common goal. Leadership provides not only a clear and shared mission and vision and a framework of robust and documented processes but also a clear and present system of risks and rewards. During Project Apollo, the risks and rewards were far greater than the monetary connotation usually associated with risk and reward. The risk was the sudden death of three astronauts. The reward was winning the space race and getting to the moon first. That set of risks and rewards is hard to match in the business world, but it begins to dismantle the common theory that annual salary increases are a reward and the threat of losing one's job is a risk. During my tenure at NASA, I was earning half of what I might have in private industry. My only fear of losing my job was no longer being part of the manned space program, not the family, social or financial consequences.

Encouraging entrepreneurial behavior at all levels is an alien concept to most business leaders. Framed in the context of Project Apollo, my definition might be more palatable. An entrepreneurial employee is anyone who has a clear, concise, and compelling mission that is all-consuming. Living the mission ensures that there is never a consideration for time clocks, overtime pay, sports, hobbies, or distractions when working on the mission. I am not suggesting being a workaholic or having a dysfunctional home life. I am suggesting that when at work, an entrepreneurial employee is preoccupied with carrying out the shared vision and mission, not with power, self-aggrandizement, or office politics. The second ingredient is a clear path to follow. There must be a road map (process procedures, standards, quality plans, reference documents, performance data) that keeps the highly motivated and focused individual traveling down the straight path to the desired destinations. There must be clear boundaries of where the ditches and hazards lie on either side of the straight road. Successful individuals will know exactly where their boundaries exist. They will accept the accountability for getting to the destination. They will learn to ask forgiveness instead of permission when they make entrepreneurial decisions that fall within the agreed-upon boundaries. Finally, an entrepreneurial employee is fully aware of the rewards for achieving success and the risks of falling short or failing. There are no such things as annual raises. Compensation is performance based on learning new skills. They are responsible to themselves and their coworkers for individual and project success. They are also acutely aware of the consequences of substandard performance. There are no coworkers who are on a free ride or unaccountable for their own actions, so all boats rise on the high tide.

One example of this behavior became clear to me on a consulting assignment at Dell Computer. A attendee was running late for our scheduled meeting. When he finally arrived, running into the building, he was out of breath and perspiring from the oppressive heat and humidity of August in Austin. He explained that he had just gone down the street and leased some office space for his group. I inquired if that was not the job of the facilities group. His response was that he needed the space that week; it was in his budget, and he would be accountable to his boss for his actions. That was a dramatic example of an employee who had a mission, shared vision, clear procedures, and the accountability to ask forgiveness instead of permission when the situation dictated.

Extrapolating the entrepreneurial employee from the spirit-de-corps and work ethic of the Apollo 13 team required the next 20 years of experience and development to present to you as valid. I have avoided the words team or teamwork because professional sports personnel are highly compensated specialists working for autocratic leaders, which is not exactly the community of entrepreneurial employees I have discovered in the Kairos moment.

History records that most idealism is short-lived. As Project Apollo gave way to Skylab and Space Shuttle, Philco became Philco-Ford, then Aeronutronic Ford, and then Ford Aerospace. Cost-cutting and quotas became more important than idealistic quality goals. The public did not hold space flight as a high priority, and NASA lost much of its funding. It lowered quality standards in proportion to the budget reductions.

13.5   WISDOM FROM KNOWLEDGE

Leaving NASA in 1980, I entered the manufacturing community with the knowledge that I had amassed from auditing hundreds of companies and from being a member of the Project Apollo team. The anecdotal and experiential information accumulated during my 14 years at the Johnson Space Center was valuable and practical for non-governmental organizations. Although I had not developed this knowledge into clinical methodology, I had the ability to analyze any given business process and offer an appropriate solution to productivity and quality problems. By using the models developed for supplier evaluation at NASA in a manufacturing company, the organization would surely make dramatic increases in output while significantly lowering scrap rates and raising profits. My thesis was valid, including the fact that most companies did not have enlightened leaders with compelling mission statements and

FIGURE 13.1 A Kairos Moment for the Author—Photo from The Author's Office on 9/11/2001

The photo is of the parting of clouds over the Sierra Nevada Mountains, creating brilliant streams of light. To the author, this was a sign that we had to learn the lessons of 9/11 and plot a bright new future from the tragedies.

endless budgets. Neither did they have cadres of trained personnel on staff awaiting the challenge of fighting fires as they ignited. Finally, private industry could not afford double and triple redundancies in its systems to avoid 99.95% of any potential disasters.

I spent the next decade as a manufacturing manager and quality manager at three manufacturing companies. In each case, my visionary leaders agreed that I would dismantle the overhead quality inspection groups, reassign them to revenue-enhancing jobs, and hold the workers accountable to the company and to each other for the quality of their own work. I was successful in each assignment.

The transition from the Kairos moment as shown in Figure 13.1 to awareness of products liability and organizational negligence was to begin as a Training Manager for a division of Schlumberger, a multi-national corporation with thousands of employees working in oil and gas exploration and drilling.

13.6 HANGIN' WITH THE BAD GUYS

In an otherwise uneventful staff meeting one day, I asked the division vice president why Schlumberger was self-insured instead of having traditional life and health insurance coverage for the employees. His explanation ("We kill six or seven people a year") astonished me and changed forever my paradigms about corporate liability.

While I had the most comprehensive mandatory safety training and awareness, many of our employees worked in the North Sea and other hostile environments where oil exploration happened. There were, unfortunately, and inevitably, several casualties that simply defied all attempts at prevention. While there was an unrelenting corporate culture to eliminate fatalities through awareness and preparation, the annual casualty rate had reached a very low but still tragic number. There were even more tragic outcomes experienced by companies with much less corporate conscience and accountability.

My rich history with NASA, the discovery of the Kairos Moment, and 50 years of experience with manufacturing and service companies have precipitated FBP, as presented in this book. I will do my best to impart learning moments and experiential anecdotes to the readers to help them maximize their outcomes in litigation.

13.7   LEADERSHIP VS. MANAGEMENT

The business lexicon typically uses the terms management and leadership synonymously. The distinction between the terms is quite profound. Some define leaders as well-respected administrators who possess well-developed interpersonal skills. Managers are defined as taskmasters, authoritarians, and, most paradoxically, benevolent dictators.

As I strive to gain and maintain a competitive edge in business, enlightened companies are moving away from classic management systems, and the terms management and leadership are taking on clearly different meanings. As I become aware of the need for a competitive edge (and immunity from liability), we need to be more exact in our definitions to avoid a communication breakdown over semantics. Further, we need to plot a course to grow from the negative traits assigned to traditional managers to the positive traits of the evolving definition of leadership.

For many generations, we have tolerated managing (directing, controlling, handling, administrating, and regulating) as a vital element of company operations. Businesspeople traditionally manage by making others submissive. They also contrive and arrange, which makes managing a noun. When it becomes a noun, it no longer adds value to the delivery system, and it becomes an overhead expense. It exists merely to control anarchy. It requires a police force. It spends much of its time reacting to adversarial behavior and sustaining itself. We have created this function called management because our family, faith, and business training all perpetuate the belief that people need to be closely watched by strong managers, or they will not work efficiently. When analyzed closely, the function includes tasks such as supervising, administering, reporting, evaluating, interpreting, decision-making, and adding little value that directly benefits the customer. We praise managers as problem solvers, although they seldom admit that anything went wrong. Their bosses applaud them for driving waste out of existing inefficient processes, although they cannot define how those inefficient processes came into being. Most management positions are non-value added, overhead positions occupied by highly talented technical experts who were promoted as a reward for tenure. They were highly trained in their chosen vocation but seldom had any skills in leading people.

Leadership does not imply the overhead that managing does. During a performance, an orchestra leader cannot manage the work of the players. He or she cannot regulate the activities of those whom they lead. The maestro stands among talented and well-trained players and leads by setting the pace and tempo and ensuring that everyone is in a locked time step. Leaders have a clear vision of their destination and a road map of how to get there. They continually share it with everyone. They drive down decision-making to its lowest level and empower the workers to act on their decisions. They allow people to fail if the failure results in training that avoids the same mistake in the future. Leaders look to their teams for ideas and processes. They do not rely on conventional wisdom, and they avoid tunnel vision. They are often visible in the workplace, coaching, mentoring, sharing, training, and adding value to the delivery system. They give direction only by logical necessity. They reward accomplishments and milestones met. They encourage continuous improvement in everyone and in every process. Leaders are constantly in training and are always being educated. They model excellence and perform every task with a high energy level. They encourage the superstars to mentor those who are having difficulty. The synergy of that relationship raises the performance of the entire team. They encourage consensus thinking but will not condone group mediocrity. They are builders of teams, models of honesty, and producers of results, not reports. When called to help with a business problem or improvement need, a competent consultant will ask for an initial meeting with the management/leadership team before proposing any solutions. The outcome of that meeting will include a list of stated issues, notes about implied issues, and bullet points about the perceptions of the people who were in the meeting. If those who attend the meeting present themselves as authoritarian, all-knowing, enmeshed in their paradigms, and fix blame on others, the chances of the consulting assignment having any true effectiveness and value are very small (managers). On the other hand, if the meeting attendees are active listeners, eager to learn, open to self-assessment, and candid in their answers, the chances of the consulting assignment being successful and providing a return on investment for the client company are much greater. A consultant should always research the company's history and any information about the senior staff before agreeing to an initial meeting. Today, that data is readily available on the internet.

What does this treatise have to do with products liability? A competent Forensic Business Pathologist will approach a case file with the exact same methodology. There is usually a wealth of data about the company and senior staff on the internet. This information begins by painting a profile of the standard of care that may be typical of the defendant. While I am often not able to interview the company leaders, the alternative is to read everything available on the company website. Marketing managers often fill the website with bragging points about the company and its accomplishments. These pages provide a wealth of information on how the company management views itself and how it deals with its customers. Over-eager web designers often include exaggerated claims and inaccurate statements on the website, which can be used in an expert's reports and depositions to make the case for their proclivity for exaggerating the truth. On one defendant's website, I found more about their stock car racing team than I did about their manufacturing capabilities. This discovery gave direction for my research that uncovered evidence of abdication of leadership by the senior staff.

If the executives are well-published and members of industry associations, the standard of care they exhibit is assumed to be at least minimally acceptable (without in-depth research). Lack of available information on the management/leadership team is also a reliable indicator that there is information that their business conduct may be less than admirable. In one case, industry articles by the "quality manager" of a company provided guidance for dealing with products liability lawsuits. That led to a stunning deposition where he revealed that he was primarily on staff to testify in the many suits brought against the company. That admission validated the fact that the standard of care of the entire management team was inappropriate and proved critical to quickly settling the case out of court.

If the organization is part of a larger conglomerate or holding company, flags can be quickly located about the mandates that the parent company may impose that can affect product quality and reliability. Another case involved a holding company that had a reputation for buying companies and imposing its corporate model of slash-and-burn management that each new subsidiary was forced to adapt. There were several articles available on the internet applauding them for taking floundering companies and quickly making them profitable. Research showed that their so-called success stories had a foundation of sending the manufacturing offshore, cheapening the product, and living on the reputation of the old company until the sub-standard products were widely detected, followed by selling the company. To wit, the Florsheim story.

There is a direct correlation between the standard of care and whether the organization has managers or leaders. Recognizing the difference between them is a key tool in the successful resolution of products liability and organizational negligence tort.

13.8   QUALITY MANAGEMENT

Still working on my engineering degree and working full-time, in 1969, NASA invited me to become one of the early pioneers in Quality Management. Out of necessity and lessons learned from the Apollo 1 fire, a new focus was emerging to prevent nonconformities that might lead to disasters before they happened. When I was an electronic technician, we had in-process inspectors, roving inspectors, final inspectors, and a government quality representative who checked up on everyone else. With its roots in the sweatshops of the early 20th century, conventional wisdom promulgated that people were bound to make errors and that the only way to detect problems and not deliver them to customers was through a series of inspection toll gates. That is, at the end of every operation (process), a third party inspected manufactured items before they passed to the next operation. At that toll gate, inspectors separated the good from the bad. The compliant components went on to the next operation. The nonconforming items were segregated and either reworked or scrapped.

While this quality control inspection process was effective, it was very expensive because the inspection steps and rework added to the labor cost, and the scrap became an expense to the bottom line. Another "expense" of quality control inspection is that it is said that it is physically impossible to detect all nonconformities, and so field failures were unavoidable. This drove industries like aerospace to build double and triple redundancies into their systems to deal with the probability of undetected defects. Unfortunately, relying on toll gate inspections is still prevalent within manufacturing companies worldwide that build defective products and deliver them to unwary consumers.

It had long been debated within the quality profession that preventing defects was much more cost-effective than detecting them after the fact. Proactivity in this endeavor would require a new discipline named Quality Assurance (QA) and another called Reliability Engineering (RE). QA pioneers launched into assessing business processes to determine where defects occurred and went undetected. Reliability Engineers looked at factors such as Mean Time Between Failures (MTBF) and Mean Time to Restore (MTTR) to predict critical paths of failures and which components were reliable or not. Today, both of those disciplines are degreed curricula in engineering schools. In the 1950s and 1960s, they were emerging areas of expertise. There were seminal works published on the correlation of process inconsistency, human error, and defect propagation through production processes.

With the consent and approval of NASA, Philco created the disciplines of Quality Control Engineering and Reliability Engineering as formal positions within the company.[4] Quality Control Engineering became an emerging engineering discipline.[3] NASA recorded our titles in the contractual documents. From 1969 to 1980, I personally authored volumes of white papers defining the role of quality assurance in assessing the qualifications of suppliers and contractors. Today, this is called "supply chain management."

In 1987, the European Union promulgated the publishing of the ISO 9000 series of quality standards as a baseline for businesses to use in determining minimal standards for quality process conformance. Today, ISO 9000 exists in more than 160 countries as the baseline standard for measuring organizational standard of care in quality management. Through my consulting work and my contributions to ASQ, my peer-reviewed publications advocate the beneficial use of ISO 9000, and I use the standard in objectively evaluating standard of care within manufacturing organizations as a foundation of FBP. I have gained esteem as an international authority on ISO 9000.

It is critical for attorneys and business executives to understand the journey from quality control to quality assurance to quality management. In products liability tort, it is not sufficient to expose the inspection processes that may have not detected a critical nonconformity. It is also not sufficient to determine if the company implemented quality assurance measures to prevent defects before they left the plant. It is imperative to investigate the quality management of defendant organizations to determine if there is a pervasive culture of quality, tools of continual improvement, and measures of performance that can assure that critical defects cannot reach the customer. Assessing the efficacy of an organization's quality management is critical to proving innocence, guilt, compensatory damages, and punitive damages. This concept is the underpinning of Forensic Business Pathology.

NOTES

1 In ancient history, learned scribes and philosophers would analyze horrific events and create parables about the lessons to be learned from these events. They became known as Kairos Moments.

2 There's controversy over whether Flight Director Kranz coined the phrase, but I heard it from him several times and it is the title of his book *Failure Is Not an Option: Mission Control From Mercury to Apollo 13 and Beyond*.

3 I am always voir dired about my degree equivalency of Quality Control Engineer and how it was granted by Philco and NASA. I explain that I pioneered seminal work in creating the discipline and the body of knowledge of quality management. I chaired the Quality Management Systems Committee of The American Society for Quality (ASQ) and contributed to the qualifications criteria for ASQ's Certified Manager of Quality and Organizational Excellence (CQM/OE) certification, which is the most prestigious professional certification in the quality discipline. I was also one of the first to obtain that certification. If that explanation is still challenged, I ask how the credentials of electrical and mechanical engineering professionals were determined before colleges instituted formal degree programs. I have never been called to a Daubert proceeding.

4 As part of our contract, we were also approved as Material Review Engineers who had the signatory authority to disposition defective materials on behalf of NASA.

Section 6

Risk Avoidance

14 The Evolution to Risk Avoidance

14.1 INTRODUCTION

My experience as a quality control engineer and a testifying expert witness has caused me to conclude that the creation of defects that lead to exposure to danger is avoidable. Risk is avoidable because it can be foreseeable.[1]

Before I had witnessed the outcome of process variability, which was catastrophic fires, injury, loss of property and death, I was not aware that our traditional approach to quality management was fundamentally flawed. Traditional preventive action is often triggered by a problem being foreseen in one or more processes. Those who employ tools such as FMEA's can uncover potentially detrimental process inconsistencies in their investigations. Those who rely on PAR's are not proactive enough to practice risk avoidance. Rarely is preventive action a result of an epiphany or flash of genius. Nor is it a result of cognitive risk awareness.

Even more uncommon is a formal program of foreseeable risk implemented as an immutable and proactive cultural mandate.

14.2 FORESEEABLE RISK

Businesses are living organisms. Each company exhibits a unique personality and has attributes distinctive to its own heritage and lifestyle. Organizations possess a measurable state of wellness that is in constant flux depending on internal fitness and the relentless assault of outside pathogens.

Business health can span the range from the peak performance of a world-class contender to being terminally ill. Overt appearance is seldom an indicator of a state of well-being at any given moment. Symptoms of the disease may be known and treated, manifest but being ignored, or undetected but growing and consuming the organism.

Foreseeable Risk is a scientific diagnostic methodology that examines the health of a business utilizing proven tools of process analysis, performance to standards, business metrics, and management system effectiveness. The first step of triage is a detailed workup of the overall health of the business backed by clinical evaluations of each constituent element of the company.

Used proactively, Foreseeable Risk is a roadmap for organizations to achieve peak health and wellness. Used as a forensic tool, it provides documented evidence of the standard of care and foreseeable risk measured against quantifiable standards of performance.

DOI: 10.1201/9781003535621-20

For enlightened business leaders, Foreseeable Risk is a strategy for achieving peak performance while immunizing the company from products liability and organizational negligence. By virtually eliminating product defects and service errors, organizations can achieve unparalleled pinnacles of customer service.

14.3   THE NEW PARADIGM

For organizations to be successful in risk avoidance, there must be a top-down paradigm shift in the vision, mission, and value system of the company. Whether your company culture has morphed organically over time or is the result of enlightened strategic planning, the concept of risk avoidance is never part of the company mission.

The fault is not one of pragmatic process development or selecting one quality management approach over another; it is embedded in the fundamental paradigms of organizational leadership. Risk avoidance is not taught in business school or in management seminars. It is not part of business or quality management training curricula.

It is a new paradigm that states that defects are avoidable, not inevitable. If your value system is based on the inevitability of some level of nonconformities being acceptable, you will always have nonconformities.

Unfortunately, since ISO 9001:2015 added *risk-based thinking* to our lexicon, quality professionals have attempted to create compliance parameters for risk-based thinking rather than taking a 50,000-foot level assessment of its meaning and implications. The following steps are not found in your father's quality management system.

1. Revisit your Vision, Mission, and Values
   a. Senior management must become fully aware and enlightened on the tenets of risk avoidance and foreseeable risk.
   b. If you are to be successful, the vision, mission, and values must be revised to include a mandate for risk avoidance
   c. The new mandates must be communicated at all levels and incorporated as a culture, not a requirement.
2. Reassess your quality management system
   a. While ISO 9001:2015-based QMS addresses business success, implementation is typically founded in the traditional steps of continually improving processes to achieve more favorable outcomes and metrics.
   b. Determine where your organization falls in Table 16.1 and determine for which level you are initially striving.
   c. The QMS reassessment process must change the focus to quality as a profit center, stupid proofing processes and products, metrics driving key business indicators, avoiding defects reaching the customer, and achieving perfect customer report cards.

    d. Process control must include inculcation of the vision, mission, and values.

    e. Each tenet of your QMS must be restated from risk management to risk avoidance guidance.

    f. Use of proven tools that have been developed for risk avoidance.

3. Implement the Risk Avoidance business model

    a. Conduct presentations and workshops to teach Foreseeable Risk

    b. Assess your compensation program. Consider a risk and reward program that has incentives for defect-free performance and penalties for creating nonconformities.

    c. Recast your version of PDCA to replace the "check" function with an "avoidance" function.

    d. Assess the control limits of your metrics to strive for zero outgoing critical nonconformities, not an acceptable number.

    e. Utilize every defect report, customer input, and corrective action as gifts for your new goal of delivering defect-free products and services.

14.4 THE TOOLS OF FORESEEABLE RISK

1. Assess and revise the steps of each process in your organization. Include two mandatory questions for the process operator to answer before moving the product or service to the next process step.

    a. Am I certain that my work is defect-free before passing it to the next process step?

    b. Am I solely accountable to the next process operator for my work product?

2. Add a step to audit and management review meetings. Ask the questions.

    a. How can the owner of the next process use my work output badly or inappropriately?

    b. How can the work output of each process affect a customer?

3. Add a step to your FMEA processes or impanel a risk avoidance committee. Include individuals who have no formal involvement in the processes but are dedicated to the success of the company. Add these steps:

    a. During DFMEA's and PFMEA's, add a round of questioning about how a customer could use the product or service badly, stupidly, or for the wrong purpose.

    b. Review customer complaints, warranty reports, and repairs with a deliberate focus on what we could have done to avoid each issue. Customer stupidity will not be a relevant factor in a lawsuit if you have not adequately anticipated misuse.

4. Continually model the culture that a defect will never leave my workstation; therefore, a defective product or service will never reach a customer.

14.5   THE TOOLS—FORENSIC INVESTIGATIONS

There are four types of forensic investigations:

1. Process compliance—Determine compliance with documented procedures.
2. Process improvement—Assess opportunities for business process excellence.
3. Risk assessment—Assess opportunities to identify foreseeable risk.
4. Expert investigation—Conduct forensic incident investigations (only for graduates of Course 5).

There are four subcategories of forensic investigators:

- Process Compliance Assessor—QM Level 2 Certification—Process Management Expert
- Process Improvement Assessor—QM Level 3 Certification—Business Process Excellence Master
- Foreseeable Risk Assessor—QM Level 4 Certification—Risk Avoidance Master
- Expert Investigator—QM Level 5 Certification—Business Management Systems Master

14.6   FORENSIC INVESTIGATION (FI)

A forensic investigation examines documented processes and their efficacy regardless of where they exist in the organization. Among the foundational tenets of the QM program is that undocumented business processes should not exist. Each process must have operational procedures and work instructions to the level of detail necessary to ensure process excellence and risk avoidance. They must also have appropriate roles and responsibilities for the process owners and operators.

Evolving processes and procedures to QM Level 1 standards is an iterative process of implementing an enterprise-wide Business Management System (BMS). Each investigation may be used as a tool to form the evolution from QMS to BMS. These investigations are more than a snapshot interview of a designated department. They are in-depth investigations using proven quality tools such as FMEA, The New Seven Tools, RCA, and creating Kaizen Events as needed for complex processes.

FI Investigations replace classic QMS audits. They include:

- Identification of the process to be investigated.
- Collection of all applicable documentation related to the process.
- Identification of Roles and Responsibilities associated with the process.
- A checklist of the planned outcomes of the process.
- Concise criteria for evaluating the compliance of the process.
- Validate process operators' competence.
- An evidence-based evaluation of each operation within the process.
- An evidence-based evaluation of each role within the process.

- Fact-based recommendations for immediate corrective action as warranted.
- Fact-based recommendations for process improvements.
- Appropriate metrics to indicate fulfillment.
- Action plans to remove foreseeable risk.

14.7   PROCESS IMPROVEMENT INVESTIGATION (PII)

PII's are designed to assess the efficacy of a process with the intent to identify opportunities for measurable improvement and advance business process excellence. One approach to consider is a small, dedicated team (one or two Quality Professionals led by a Business Success Champion) that pulls in internal functional experts for a short period to conduct a deep dive into a specific business process. A Kaizen sprint can be used to find weaknesses and potential risks that need to be avoided and then update the process based on the sprint findings. This reinforces the iterative processes previously mentioned in FI.

These investigations include:

- Identification of the process to be investigated.
- Collection of all applicable documentation related to the process.
- Identification of Roles and Responsibilities associated with the process.
- Current metrics of process effectiveness against business (functional level) Key Process Indicators.
- Current competency evaluations of the process owners and operators.
- An evidence-based evaluation of each operation within the process.
- An evidence-based evaluation of each role within the process.
- Specific process improvement recommendations with anticipated enhancements in metrics.
- PII follow-up metrics monitoring at appropriate intervals.

14.8   RISK ASSESSMENT INVESTIGATION (RAI)

The RAI should not be confused with risk management programs that may be implemented because of statutory or regulatory requirements or levied by insurance carriers. Organizations that have implemented risk management programs to ISO 31000:2018 may have invested considerable time and effort attempting to comply with ISO 31000 and may be mired in the complexities of that Standard. The point is NOT to add another compliance standard to an organization's infrastructure.

First, RA Investigations begin at the process level. Each process must be scrutinized in rigorous detail. We have many tools in our quiver that we can use diagnostically, such as DFMEA's, PFMEA's House of Quality, Ishikawa Diagrams, etc. There is no need to construct complex matrices of risk management to diagnose foreseeable risk in a process or begin new risk avoidance initiatives. When I have completed an RAI on each process, the operational procedures, work instructions, metrics, and roles must be revised to ensure the investigation results are implemented. As I link individual processes that are immune from foreseeable risk, I grow a tree that eventually addresses all areas of organizational risk. RAI's must start at the lowest level in the production process. Following are elements of a good RAI:

- Identification of the process to be investigated.
- Identification of interrelated sibling processes that can be affected by associated risks.
- Gathering of historical audits and Findings as a baseline for the current audit. Are there any repeat Findings? If so, were associated risks properly defined?
- Collection of all applicable documentation related to the process.
- Identification of Roles and Responsibilities associated with the process.
- Identification of any risks previously discovered in the process.
- Identification of all potential risks that the process can generate either directly or indirectly.
- Current competency evaluations of the process owners and operators.
- An evidence-based evaluation based on "what could possibly go wrong?"
- An evaluation of how customers can potentially misuse the product or service.
- Identification of potential foreseeable risk within the definitions of duty of care and standard of care.
- Specific process improvements that will eliminate the root cause of the foreseeable risk.
- Follow-up metrics monitoring at appropriate intervals.

RAIs are part of a perpetual enterprise-wide program of identifying foreseeable risk. They can also be used to diagnose risks as shown in Table 14.1 within statutory and regulatory processes.

TABLE 14.1
ISO 9001:2015 And The BMS Model

ISO 9001:2015 REQUIREMENTS	Organizational Excellence
3 Terms and definitions	**Define Quality as a Profit Center**
4 Context of the organization	**Enterprise Excellence**
4.1 Understanding the organization and its context	**Creating an Enterprise-Wide Business Management System (BMS)**
4.2 Understanding the needs and expectations of interested parties	
4.3 Determining the scope of the quality management system	
4.4 Quality management system and its processes	
5 Leadership	**Enlightened Leadership by Example**
5.1 Leadership and commitment	**Create Vision, Mission and Values Establish Roles, Responsibilities and Metrics**
5.2 Policy	
5.3 Organizational roles responsibilities and authorities	

TABLE 14.1 *(Continued)*
ISO 9001:2015 And The BMS Model

ISO 9001:2015 REQUIREMENTS	Organizational Excellence
6 Planning	**Planning the BMS**
6.1 Actions to address risks and opportunities	**Define Objectives, Processes, Change Control**
6.2 Quality objectives and planning to achieve them	
6.3 Planning of changes	
7 Support	**Outcome-based Risk and Reward**
7.1 Resources	**The Responsibilities of the BMS Implementation and Maintenance Team**
7.2 Competence	
7.3 Awareness	
7.4 Communication	
7.5 Documented information	
8 Operation	**The Process Based Organization**
8.1 Operational planning and control	**Creating Value-Add Processes Removing Opportunities for Nonconformities**
8.2 Requirements for products and services	
8.3 Design and development of products and services	
8.4 Control of externally provided processes, products, and services	
8.5 Production and service provision	
8.6 Release of products and services	
8.7 Control of nonconforming outputs	
9 Performance evaluation	**Evaluating Effectiveness**
9.1 Monitoring measurement analysis and evaluation	**Investigations and Metrics**
9.2 Internal audit	
9.3 Management review	
10 Improvement	**Improving Effectiveness**
10.1 General	**Maintenance of the BMS**
10.2 Nonconformity and corrective action	
10.3 Continual improvement	

This table lists the clauses of ISO 9001:2015 on the left and then defines the BMS 9001:2024 definition of conformance and performance and of risk management to risk avoidance.

14.9   EXPERT INVESTIGATION (EI)

An expert investigation is the ultimate in problem-solving techniques. It is the tool of choice when examining a defendant company in a lawsuit to determine if they exhibited an appropriate or negligent standard of care in introducing their products into the stream of commerce. Simply stated, when a product fails and causes injury or harm. An in-depth investigation is mandatory to determine if the manufacturer sold a defective product or one that was not fit for use. The plaintiff in the case will have filed a suit that claims injury and/or harm that was the responsibility of the defendant, the manufacturer.

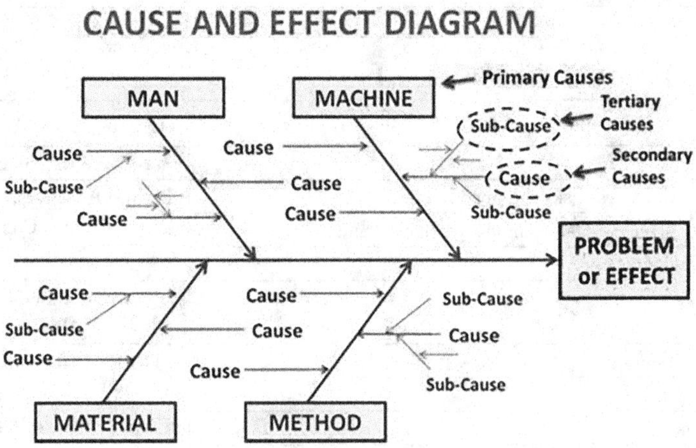

CAUSE AND EFFECT DIAGRAM

FIGURE 14.1   Ishikawa method for forensic investigations

Karou Ishikawa designed this very helpful methodology for graphically representing cause and effect in solving problems.

An EI digs much deeper than a traditional FMEA and is more thorough than a root cause analysis as depicted in Figure 14.1. It makes a case that will stand up to examination by defense attorneys in a court of law, to challenges of facts, to examination of methods, and to argue the opinions offered. An organization does not need to wait to be sued to do an EI. If your products are inherently dangerous or your customers can easily misuse or abuse your products, then conducting an EI is certainly worth the time and effort.

14.10   THE BALANCED SCORECARD (BSC)

A Balanced Scorecard (BSC) is a strategic management performance metric used to identify and improve various business functions and their outcomes. It measures past performance data and provides organizations with feedback on making future business decisions.

A BSC can be used as a standalone tool for management reviews or to summarize the results of a Forensic Investigation. After you have conducted your process compliance, improvement, and foreseeable risk investigations and have identified or updated your business and functional KPIs, a scorecard can be used to track your progress toward achieving your Vision, Mission, and Goals.

A BSC looks at much more than process efficiency or product defect rates (see Figure 14.2). It allows you to show structure within your overall company strategy. While each internal department may track individual KPIs differently, a BSC allows management to organize these in an easy-to-understand format. Management can see where they are now, where they want to be in the future, and how they plan to get

FIGURE 14.2 Depiction of a Balanced Scorecard

A Balanced Scorecard is one of many tools available to quality professionals. It gives a map of how the organization is functioning internally and is also a method to determine customer satisfaction.

there. It is flexible enough for even the smallest company to create and can be used by companies that have multiple sites for comparison. The key to a successful BSC is in the last column—the frequency check. Realistically reviewing your status on a periodic basis forces you to adjust on a more regular basis, which in turn helps you stay healthy in an ever-changing business environment.

NOTE

1 Assumed risk is when individuals are involved in new technology with multiple variables, such as being a test pilot. They acknowledge ahead of time that their activities can lead to tragic outcomes. Foreseeable risk is when processes are established to repeatedly create conforming products or services, and, with the proper tools and training, deviations can be avoided.

Section 7

Process Excellence

15 The Evolution to Process Excellence

15.1 INTRODUCTION

The first step is obtaining full backing from senior management that the proposed Business Process Excellence (BPE) project starts by transitioning the QMS to a Business Management System (BMS). The proposal you present[1] must be convincing, not with overstated promises, but with specific, measurable objectives. The discovery and mapping phase of the project will be extensive and require the time of people who have jobs to do. The proven way to gain their support is from the famous radio station WII-FM; what's in it for me.

Unless everyone you pitch sees a direct result that improves their performance or causes them less work, you will not garner their complete cooperation. You will be identifying, mapping, and analyzing each process. Every time you do that, you are likely to find some inefficiency in a process that yields an immediate, measurable benefit. When you include identifying foreseeable risk in this process, you have a good chance of removing some activity that is detrimental to the organization.

The next step is creating the organizational map that shows the input to each process and the output. This creates a supplier and a customer for each process. As you get into the nuances of BPE, you will discover that by creating a supplier-customer and a customer relationship for each process, these groups will begin talking to each other to air problems that they may have had for protracted time periods. In one machine shop, the downside was that customers for each process had the chance to modify how many defects were produced upstream. A mill operator could hide their own defects by cooking the numbers from the raw stock provider.

In the past, this type of problem could have been addressed with punitive actions. In this case, the issue was solved by having the upstream supplier move the product to the next workstation, hand the traveler to the next process owner, and have them initial the quantity they received. It wasn't long before the supplier started to understand the needs of the customer, and they worked together to make the processes robust. Since no one wanted to be accountable for producing scrap, the defect rate also fell precipitously.

This organizational process map (see example in Figure 15.1) will be among the most challenging deliverables you and your group ever devise and maintain. Yes, I suggest forming a standing committee that is responsible for developing and continually collecting data from the process map. These committees often have rotating membership, but they must include representatives of each department who have skin in the game of business process excellence.

DOI: 10.1201/9781003535621-22

FIGURE 15.1 The Process Excellence Model

This is a graphic representation of the new process model described in this book. It is meant to compare to the classic process model depicted in Figure 12.7 where the process operator is now responsible for no defects moving from one work center to another.

15.2   BEGINNING THE JOURNEY

It is common to use internal auditing to begin assessing an improvement project. A common definition for the globally accepted scheme of compliance audits is a systematic, independent, and documented process for obtaining objective evidence and evaluating it objectively to determine the extent to which the audit criteria are fulfilled. In the case of an audit of an ISO 9001:2015 QMS, the audit criteria are the Standard, operational procedures, and subordinate QMS documentation.

As defined, classic QMS internal auditing is restricted in scope to the QMS and doesd not include the overall business operations. Third-party QMS auditing is even more prescriptive since the auditing agency typically provides a written scope of the audit and must stay within that scope.

The output of these audits is most often simply an audit report. This report typically cites what was observed compared to the governing procedures. The audit checklist used to create the audit report is often filed as background information and not included in the report. Within the report are statements regarding the suitability of a process compared to its objectives. These statements are typically more opinion than forensic evidence.

The auditor is then free to make comments on the effectiveness of the process against their own training and experiences. The auditor can add any observations for corrective action or improvement opportunities. Again, these are typically experiential and not compliance-oriented.

The most effective tool to begin a BPE project is a forensic investigation.

15.3   THE MASTER PLAN

Once these investigations are complete, the data needs to be verified, validated, and published. Any immediate corrective action must be done using your existing tools.

I will not propose any template for a master plan. This process must already be in your wheelhouse or that of your team members. As an internal champion or consultant, the plan should be of a known effective format. It must include:

The steps of implementation

- Who does what
- A schedule
- A budget
- A reporting mechanism
- A modification processes
- Business Metrics
- A perpetuity plan

Again, we are not suggesting how to do this because it is different in each business culture. I just propose that it must be done to avoid risk and achieve business process excellence. Whether or not your organization will take the challenge to achieve the goal of no defect ever reaching a customer is a gauntlet I throw down to the owners, not to you.

NOTE

1 This training is for quality Professionals who want to become quality champions and masters of driving risk from an organization. It will likely become your role to propose BMS to others in your company and create a unified plan to present management.

Section 8

The Genesis of The New
Quality Professions

16 The New Quality Professions

16.1 THE IMPERATIVE

The imperative is to evolve traditional QMS into a business management system (BMS). This is the call to action to recast quality policy and objectives into vision, mission, and values that senior management must identify as the immutable core imperatives of their business.

In defining the processes and their interactions, include the tenets of risk avoidance in the processes. I will discuss later the benefits of developing the workflow into a culture of accountability among the process owners and operators.

16.2 ELEVATE QMS TO BMS

Those of us who implement and maintain quality management systems are typically focused on conformance and continual improvement. In achieving ISO 9001 certification, we have proven that our QMS meets the requirements of the Standard. We perform periodic audits to ensure ongoing conformance to the Standard. In between internal and external audits, we shift our focus to preventive action and continual improvement.

We are proficient in the tools of process improvement such as Six Sigma, Lean, 5S, Kaizen, Kanban, DMACR, Poka-yoke, DFMEA, PFMEA, and RCA. Effective quality leaders will use these tools to solve problems, minimize defects, and hone process effectiveness. Unfortunately, senior management seldom shares our passion with the tools of quality management. Their prime focus is typically on increasing profits and market share. If you are part of a public corporation, there is an additional demand on leadership to meet key performance indicators set forth by a board of directors. Fundamentally, we don't even speak the same language.

Too often, quality system management is an overhead expense that is a necessary evil to avoid risk and prevent defects from reaching customers. The premise is that senior management creating the quality policy and objectives and attending management review meetings are often ISO 9001 conformance exercises rather than integral to their business objectives. In other words, we coexist rather than being collaborators.

Quality professionals are often unaware that our tools define us. We become so invested in the business of quality that we do not see the big picture of how the organization functions at the macro level. Our tools and perspective as the champions of quality become who we are. WE observe Six Sigma Black Belts, who are so devoted to the culture of what they do that they lose sight of the key objectives of the organization as a business entity.

DOI: 10.1201/9781003535621-24

To transition from quality management to business management, we may need addiction intervention to begin the journey to becoming business-focused. While our tools serve us well, they are limited in their scope to process improvement, not to the prime directive of the organization as a complete living organism.

This book challenges quality professionals to create a new paradigm about what we do and why we do it. It will create the foundation for evolving quality management into business management. It will encourage the harmonization of the goals of senior management and those of continual improvement and risk avoidance.

It makes the case for you to evolve the requirements of ISO 9001 into processes that are value-added rather than conformance goals. The objective of this transition plan is to become champions of helping senior management advance their key performance indicators instead of just improving quality metrics. Implemented effectively, quality as a profit center will become an integral tool of business success instead of an overhead expense of process management.

The first step is the transition from being a quality advocate to becoming a business success champion. We often believe that, as part of implementing Clauses 4, 5, and 6 of the Standard, we will be collaborating with leadership on creating a new lexicon and mission based on the Standard. In implementation meetings where we discuss interested parties, policies, roles, responsibilities, and objectives we attempt to educate leadership in our language and culture.

Unfortunately, the more we are focused on the Standard and tools of quality management, the wider the chasm becomes with the needed leadership commitment. Those who will be most successful at making PDCA part of the culture of their organization will realize that they must take proactive steps to learn how to communicate with senior management.

Here are some tips for beginning the journey to creating a genuine dialogue with senior management.

- Visit your company website and discover what the organization is communicating to the world about what you do and who you are.
- Get a copy of the annual report and become educated on what is important to senior management.
- Compare what you learn to what objectives you currently have as priorities.
- Once you have learned business-speak and begun adopting the lexicon of administration, find a member of management who may be a potential champion for you. Practice communicating the benefits of a formal management system in the vocabulary of business management.
- Then, ask your champion to request a series of meetings with senior management to revisit the policy and leadership role with the objective of creating a shared vision, mission, and value set. Focus on the key business indicators that are critical to leadership, and transition quality metrics to business metrics.
- Look for every opportunity to create a return on investment for every element of the Standard and the QMS. Particularly in ISO 9001:2015, within clauses 7 through 10, you must find how to demonstrate that compliance with a requirement has a measurable benefit to business objectives and to customer satisfaction.

16.3   SUNSETTING THE QUALITY POLICY

A traditional QMS contains a quality policy. That seldom reflects the vision and values of the leadership team. Most companies have a management staff, but true leadership is not part of the culture. There are typically operational procedures, job descriptions, and an employee handbook, but seldom do organizations define boundaries of behavior. Finally, the point I have been making all along is that the only metrics that matter are business metrics that reflect how well I've achieved our vision and mission.

For the rest of this section, I am going to compare the nine tenets of BMS to how we won the space race and landed men on the moon. In my years of consulting work, I have never found a CEO who had a greater challenge than Project Apollo, so I will use that as a benchmark of how you will be implementing your BMS. This chart is what I call the Seven A's. They elevate the traditional PDCA to a process excellence plan.

16.4   VISION

I was in high school when President Kennedy announced his vision to land men on the moon by 1970. He was talking to me. On my 19th birthday, I packed my car and moved from New York to Houston. I enrolled in the University, got a job to support myself, and pestered enough people so I landed an electronics technician job with Philco, the prime contractor building the Mission Control Center. That is the power of a compelling shared vision. The vision for your organization must come from the leader and be a cultural imperative for everyone (Table 16.1). Suggestions for your vision include:

a. Be the brand leader
b. Critical defects never reach a customer
c. Industry's highest customer satisfaction
d. Model workplace safety
e. DO NO HARM

16.5   MISSION

The mission plan defines how we will achieve our vision. It is the master plan for all to follow. In your organization, the mission may be stated as we provide products and services of such exceptional value, reliability, and safety that customers will demand them. Also, success is inevitable and sustainable.

16.6   VALUES

For Project Apollo, lead flight director Gene Kranz set the value system for Mission Control. His famous declaration that failure is not an option was ingrained into our every action at work. A suggested value set for your organization might include:

• We all grow and prosper together based on our individual and group accomplishments.

Success comes from providing products and services of intrinsic value.

TABLE 16.1
What is your Organization's Vision?

	Attributes	Perceived Value	Adherence to Specifications and Standards	Reliability	Aesthetics	Human Traits	Organizational Traits	Rank Level
	Of the finest in the world	Priceless	Cannot be imitated	Generational	Appealing to the very elite	Sainthood candidate	Beyond reproach over time	1
	Best in class	Extravagant	Sets the standard	Lifetime	Appreciated by the most discerning	Sets the standard of behavior	Sets organizational excellence	2
Positive	Excellent	Rewarding	Performs well above	Decades	Provides a "wow" for most	Role model	Consistently superior	3
	Superior	Above average	Exceeds	Years	Elegant	Leadership figure	Highly effective	4
	Acceptable	Average	Meets	Low maintenance	Admired by most	Consistently positive	Consistent in output	5
	Mediocre	Low value	Barely meets	Frequent maintenance	Invisible to most	Inconsistent	Inconsistent	6
	Barely acceptable	Cheap	Inconsistently meets	Continual maintenance	Disdainful to most	More positive than negative	Unpredictable	7
	Barely unacceptable	Bargain	Inconsistently falls below	Unreliable	Ugly	More negative than positive	Dysfunctional	8
Negative	Substandard	No value	Consistently below	Unworkable	Trashy	Untrustworthy	Adversarial	9
	High-risk	Negative value	Never	None	None	Immoral	Devious	10
	Catastrophic	Painful	Performs adversely	None	None	Criminal	Criminal	

This table is a representation of the corporate behavior of companies from the very best to the criminal worst.

16.7   LEADERSHIP

Your success depends on knowing the difference between management and leadership and acting upon it.

Each day was a maelstrom of challenges, deadlines, anticipation, and precision work as we were building the subsystems for NASA's Mission Control Center. We had taken over controlling manned spaceflight with Gemini 4, but there was much new hardware and software needed for controlling Project Apollo.

We weren't aware of it at the time, but there was an ingeniously controlled chaos going on among the thousands of contractors and government workers who had one common goal: to win the space race. President Kennedy challenged us with the vision. NASA planned the mission. We executed the vision and mission.

For the last ten of the 14 years I worked at Mission Control, I was part of a new discipline of quality control engineering. Just as I had to invent new technology, we invented new roles and responsibilities needed to ensure mission success. My group's task was to design quality and reliability in the ground control systems.

At age 25, I was an active participant in critical design review meetings. I performed supplier qualification audits at companies as large as Digital Equipment and as small as two-person machine shops. I witnessed acceptance testing and was the final sign-off before systems were shipped to NASA and before they were commissioned as mission-ready.

How was I able to do all those things as a mid-level employee of an aerospace contractor?

16.8   REALITY SETS IN

I have spent the last four-plus decades studying how we accomplished landing men on the moon in seven years. I have worked with more than 700 organizations in that time and continually searched for why companies so often fall short of their intended goals.

The common thread is that senior management was seldom mission-driven. They were obsessed with profit, market share, and output. In all but a very small number of organizations I have studied, the employees managed to accomplish tasks, not to fulfill a vision.

It wasn't until my last corporate position, manager of training for a division of a large multinational organization, that I was able to coalesce and codify the environment and processes I created in building America's space program. I was tasked with creating self-directed work teams. My group created amazing results, especially with a workforce of European, Asian, and Latin personnel.

In my first twelve books, I have documented the journey of attempting to help business leaders create robust business models based on what I call The Apollo Business Model™. The plan is easy to understand but incredibly difficult to implement effectively.

16.9   DEFINING LEADERSHIP

The challenge is to find senior businesspeople who understand the difference between management and leadership. Leaders create the vision, mission, and value sets that allow their people to invent, create, and accomplish impossible tasks. Managers create an environment of command and control. They attempt to achieve their goals through power and intimidation.

If the early NASA pioneers had been command and control managers, we would have never gotten to the moon. The President created the vision of landing men on the moon before 1970. The NASA leadership created the mission plan and provided the resources needed to accomplish the vision. They created a work environment where we could all be creative and contribute our unique skills to mission success while building equipment that was 99.95% reliable.

They were leaders, facilitators, communicators, and listeners. They created the environment where we volitionally worked endless hours to meet launch deadlines and then jointly celebrated success after each mission with "splashdown parties."

When business managers took over NASA operations, mission successes were the result of dedicated aerospace professionals doing their jobs for the satisfaction of their work outcome. The downside to the lack of enlightened leadership first manifested itself with the Challenger Disaster.

Management gave the go to launch a mission at temperatures below launch parameters to save money on de-fueling the rocket and to maintain schedules on which they were graded. Seven souls died, and the Shuttle program was shut down for 32 months because we had managers instead of leaders.

16.10   THE FIRST STEPS TO ENLIGHTENED LEADERSHIP

While those who run organizations will never have the opportunity President Kennedy had to challenge an entire nation to achieve the goal of reaching the moon, those who will become true leaders must identify and codify their vision. That vision must be clear, concise, action-driven, and have defined outcomes. It must be shared and lived by everyone.

From that vision, an enlightened leader builds an inner circle to create a mission plan to implement the vision. The mission plan defines objectives, resources, milestones, and measurements of mission success. That plan is implemented enterprise-wide. The processes are continually measured, and the plan is modified as conditions dictate.

Finally, the true leader identifies the values that he or she holds as immutable and makes those values integral to the organization. The values manifest themselves in how each person is respected, appreciated, and compensated. They also create a set of boundaries from which moral, ethical, and legal values are never compromised.

16.11   ARE YOU A CANDIDATE?

Leadership is a learned skill. Creating a healthy, learning, growing, and successful organization is achievable. Your challenge is to identify your barriers to success and to create the vision to overcome them.

If you are controlled by a board of directors, you may never be able to realize the potential of true leadership. If you are an entrepreneur, you may never take the time to create the vision, mission, and values and work as hard on your business as you do in your business.

The concepts are clear. There are mission plans that can help you achieve your goals. The question is, are you dedicated to becoming an enlightened leader, or are your challenges more difficult than fixing the Apollo 13 spacecraft 200,00 miles from Earth with spare parts and duct tape?

16.12   PROCESS

At Mission Control, there was a process plan for everything. The standard compels us to use the process approach in our QMS. To create a BMS, the process approach is enhanced to include:

a. Documented processes
b. Virtual elimination of variability
c. Adherence to the plan
d. Continual improvement
e. Risk avoidance

Again, I have enhanced our classic understanding of the process model to add more detail as to what each step entails. It is intended to help the transition from QMS to BMS.

16.13   BOUNDARIES

Most dysfunctional organizations have no defined boundaries of behavior and immutable rules of conduct. A flow chart was copied from a NASA training manual circa 1966. It created rudimentary but highly effective rules of the road for how missions were planned and executed.

For your organization to achieve business excellence, there must be one set of boundaries for everyone. Our rules of the road need to be:

- Define "always" and "nevers." The always are behaviors that we do every day. The nevers are behaviors that are a dischargeable offense.
- Define the consequences of unacceptable behavior. There must be clear and consistent consequences of stepping outside the boundaries.
- Live consistently within the boundaries.

No exceptions, no excuses!

16.14   METRICS

At Mission Control, we measured everything. Meetings were held daily to analyze the data and recommend corrective or preventive measures. The data was provided to the senior staff to act upon. Failure was not an option. If you can't measure it, don't do it!

In your new BMS, metrics include:

- Every process has defined measures of performance.
- Data is gathered consistently.
- Performance parameters and limits are established and are mandatory.
- All business decisions are based on proven data.
- Processes are improved based on metrics.

Also, the prime directive is that all metrics are tied to the organization's KPIs[1] and performance goals.

16.15   CONSISTENCY

We flew 17 Apollo missions with no loss of life.[2] Each mission achieved its mission goals, except for Apollo 13. The explosion in the service module was the result of human error on the ground. Living through the daily challenges that kept occurring in space gave me the grounding for following the process model for all events. Each fix that they made in space had documented procedures that were tested on the ground and implemented by the book.

Consistency in your organization must include the following.

- Processes that are repeatable and effective achieve quality, reliability, safety goals and profitability concurrently.
- Consistency is not boring; it is the foundation for creativity and growth.
- Learn from failure. Turn problems into opportunities to improve.

16.16   ACHIEVEMENT

Landing men on the moon has been the vision of most civilizations throughout time. It is likely that your anticipated achievements are not this grand. Our new objectives to colonize space are much more ambitious and have much greater risk than lunar landing.

To conclude the space metaphor, I look at each project as a mission that employs the tenets of the BMS model. They include:

- Celebrate the success of each mission.
- Capture the lessons learned from each event and use them for corrective or preventive actions.
- Reward accomplishment, not activity. Since the advent of direct deposit employees, there is no sense of the value of compensation.
- Create the success model for the next mission.

NOTES

1 Key Process Indicators.
2 The Apollo 1 fire was on the launch pad, not in space.

Appendix A: The Quality Masters' Program

THE NEW QUALITY PROFESSIONS

	CERTIFICATION	BADGE	PRE-REQ	NEW SKILLS	CASE STUDY	CAREER OPPORTUNITY
1	Certified Quality Practitioner	CQP	2 years' experience in implementing and operating quality management systems	Demonstrated competence in QMS deployment and operations. Concept of transition from QMS to Business Management Systems (BMS)	At least one improvement project of evolving QMS from corrective action to performance	Operations
2	Certified Process Expert	CPE	Completion of Course 1 and improvement project	Demonstrated competence in evolving from QMS conformance and compliance to robust process control	At least one improvement project of evolving QM process improvement	Supervisory
3	Business Process Excellence Master	BPXM	Completion of Course 1-2 and improvement project	Mastery of Business Process Excellence. Facilitator, trainer, implementer	At least one improvement project evolving process control to excellence	Managerial
4	Business Risk Avoidance Master	BRAM	Completion of Course 1-3 and improvement project	Mastery in transitioning traditional risk management to risk avoidance	At least one improvement project evolving risk management to avoidance	Senior Management
5	Business Management Systems Master	BMSM	Completion of Course 1-4 and Case Studies	Mastery of evolving industry from conformance to performance. Teacher, facilitator, consultant	Industry leadership in BMS vs QMS	Consultant

FIGURE A1.1   The New Quality Professions

This table represents the Quality Masters Training and Certification Program and what composes each level of the five certification courses that are described in Appendix A1.1.

The Quality Masters' Training and Certification Program

AN EXEMPLAR GLOBAL RECOGNIZED TRAINING PROVIDER

A QUALITY DIGEST STRATEGIC PARTNER

A TRAINING AND CERTIFICATION AUTHORITY THAT DEFINES THE
FUTURE CAREERS OF QUALITY PROFESSIONALS AND THEIR ROLES
IN BUSINESS EXCELLENCE AND RISK AVOIDANCE.

The Quality Masters' Training and Certification Program (QM) is a structured learning system that provides an ever-advancing career path for quality professionals. It builds on the rich history of the founding fathers. It incorporates all the acquired skills and industry experiences of the candidates. It reacts to the changing needs of business and technology by casting quality professionals as champions of business success rather than as overhead gatekeepers of conformance.

Since the Industrial Revolution, the pioneers of quality have walked beside the titans of industry to ensure that their products could be manufactured with minimum defects and shield the customers from harmful nonconformities. From the earliest forms of process control, culling bad products from good, we have worked tirelessly to create more proactive systems until we currently are apparently content with minimizing defects to a statistically acceptable level.

As such, our roles will never move from reactive to a positive position on the balance sheet until we change our fundamental paradigms. We must evolve from the gatekeepers of quality to the champions of business process excellence and risk avoidance.

QM is a five-tiered study and lab approach to defining our levels of expertise and our ability to lead the changes demanded by industry and technology. There are exams throughout the lessons to ensure comprehension. Upon completion, students must provide evidence that they have mastered the courseware by being a positive influence on their existing company infrastructure. Without creating new overhead exercises, they tackle existing problems and challenges and demonstrate how their QM training was applied for measurable process improvement.[1] A successful improvement project is required to move forward to the next course and earn a certificate and badge.

Moving through the five levels of certification is at the student's pace, but they must demonstrate implementation mastery they attained at each level. The highest level of demonstrable achievement might be evolving a QMS into an enterprise-wide Business Management System (BMS), but this is only attainable with Level 5 competency. Realistically, this is a two-year minimum timeframe. The student receives professional recognition while the organization realizes new heights of process excellence, risk avoidance, and measurable financial success.

THE QUALITY MASTER'S PROGRAM

Certified Quality Practitioner (CQP)

1.1 Prerequisites—The candidate must have a minimum of two years working within a quality management system and be skilled in the deployment of

ISO 9001:2015 or another of the harmonized standards. Candidates must have an innate desire to move from ensuring quality conformance to maximizing their positive effect on business outcomes.

1.2 New Skills—Course 1 takes you through the history of the quality professions and of how practitioners are often exposed to skills training and on-the-job experience in defect detection and prevention. It provides a challenge to evolve from being stewards of conformance to champions of process excellence and to break the paradigms that sub-optimize our work.

1.3 Case Studies—After completing the online courseware and passing the lesson exams, the candidate will look within their own culture and find opportunities for using their new skills to create continual improvement projects that will be presented as case studies. These case studies will provide before-and-after metrics, new Key Processes Indicators, and/or other projects showing process improvements. Upon verification of course mastery, the certificate and badge will be issued.

1.4 Career Advancement—Attaining the CQP certification prepares the graduate to apply for progressively advancing positions within their organization or in new opportunities involving growing responsibility. These are typically operational-level positions. Graduates will bring substantial value-added knowledge and valuable new skills to any existing QMS.

Certified Process Expert (CPE)

1.1 Prerequisites—The candidate must have successfully completed Course 1 and the continual improvement project.

1.2 New Skills—Course 2 begins the journey from moving beyond the tools and norms of quality management to opening the door to organization-wide business process management (BMS). It highlights the wasteful initiatives of focusing on process conformance. It replaces them with the value-add of proactive business process performance as a tool for the elimination of process variability and waste.

1.3 Case Studies—After completing the online courseware and passing the lesson exams, the candidate will look within their own culture and find opportunities for using their new skills to create continual improvement projects that will be presented as case studies. Course 2 will be most effective if case studies focus on identifying the wasteful processes within a QMS or streamlining the interactions between up and downstream processes.

1.4 Career Advancement—Attaining the CPE certification prepares the graduate to apply for progressively advancing positions within their organization or in new opportunities involving growing responsibility. These are typically operational-level positions. Graduates will bring substantial value-added knowledge and valuable new skills to any existing QMS.

Business Process Excellence Master (BPXM)

1.1 Prerequisites—The candidate must have successfully completed Courses 1–2 and the continual improvement projects.

1.2 New Skills—Course 3 is the first mastery level of QM. It is the transition from process conformance to business process excellence. It uses the structure of ISO 9001:2015 as a template for creating a robust BMS.

1.3 Case Studies—After completing the online courseware and passing the lesson exams, the candidate will look within their own culture and find opportunities for using their new skills to create continual improvement projects. Course 3 will be most effective if case studies focus on the transition from QMS to BMS.[2]

1.4 Career Advancement—Attaining the BPEM certification prepares the graduate to apply for progressively advancing positions within their organization or in new opportunities involving growing responsibility. These are typically senior operational or enterprise-wide level positions. Graduates will bring substantial value-added knowledge and valuable new skills to existing Integrated QMS.

Business Risk Avoidance Master (BRAM)

1.1 Prerequisites—The candidate must have successfully completed Courses 1, 2, 3 and the continual improvement project.

1.2 New Skills—Course 4 is the second mastery level of QM. It is the transition from traditional risk management and risk-based thinking to risk avoidance. It uses the structure of ISO 9001:2015 as a template for creating a robust BMS.

1.3 Case Studies—After completing the online courseware and passing the lesson exams, the candidate will look within their own culture and find opportunities for using their new skills to create continual improvement projects. Course 4 will be most effective if case studies focus on the transition from risk abatement to eliminating the causes of risk.

1.4 Career Advancement—Attaining the BRAM certification prepares the graduate to apply for progressively advancing positions within their organization or in new opportunities involving growing responsibility. These are typically operational-level positions and often integrate into different functional areas to streamline interrelated processes. Graduates will bring substantial value-added knowledge and valuable new skills to any existing QMS or risk management efforts.

Business Management Systems Master (BMSM)

1.1 Prerequisites—The candidate must have successfully completed Courses 1–4 and their continual improvement projects.

1.2 New Skills—Course 5 is the culmination of QM training. Graduates will be mastery-level implementers, facilitators, trainers, and consultants in business process excellence and risk avoidance. They will be the leaders within the quality professions.

1.3 Case Studies—After completing the online courseware and passing the lesson exams, the candidate will look within their own culture and create seminal opportunities that drive business success with cultural evolution.

They will provide the breakthrough tools to help organizations achieve their highest level of success.

1.4 Career Advancement—Attaining the BMSM certification prepares the graduate to lead a BMS and to expand horizons to become consultants and subject matter experts (SME)

Section 1   Student Projects

EACH COURSE IN THE QM PROGRAM REQUIRES A PROJECT BE COMPLETED BEFORE TAKING THE NEXT COURSE. THIS IS AN EXAMPLE OF A STUDENT PROJECT.

Student Project for Course 1 (Requires Submission and Acceptance Before a certificate and badge can be awarded)

In Course 1:

- We gave you a refresher course on the history of quality.
- We looked at who we are and what our skill sets are.
- We created the model of how organizations evolve like people.
- We went deeper and looked at the pathology of organizations.
- We then related the common pathogens to the clauses of ISO 9001:2015.
- We did a deep dive into the folly of conventional risk management and risk-based thinking.
- Finally, we set the stage for the paradigm shift that must occur in management systems to move beyond conformance to performance.

Before you are eligible to take Course 2, you must develop case studies, excerpts of current QMS and improvement programs, an audit of your current QMS, or some other form of written proof that you have taken the lessons of Course 1 and used them to assess the current state of your QMS and organization.

This project should not be an assignment that consumes overhead hours. Instead, it needs to evolve over months of you performing your job and taking notes of how your organization differs from the content of Course 1. The key here is taking the lessons learned and showing how you apply them in your day-to-day work.

You can format the report using any method you are comfortable with. We will provide you with a URL to upload it. MS Word or Google Docs are the preferred programs. Two Senior Quality Masters will review your submission and evaluate how well you internalized the course and how skillfully you applied the lessons in your organization. There is no pass or fail. If we conclude that you did not absorb the training, we will require you to take all or part of the course again at no additional changes.

Just as the case studies provided are part of the quizzes, they are also part of the evaluation. We suggest that you find the opportunity to cite how these case studies affected your championing the continual improvement of your organization.

When you become a Certified Quality Practitioner, you will be ready for Course 2, Certified Process Expert. In this lesson, you will utilize the refresher course in Lesson 1 to begin the process of migrating from a quality management system to a whole-enterprise business management system. You will learn how to transform a quality policy with a vision, mission, and values that are created by senior management, not the quality department. There is a tutorial to help you develop your leadership skills. We will help you overcome your paradigms about conformance assurance to focus on organizational excellence. We will reinforce how to kick-start a business management system by using ISO 9001:2015 as the platform.

When you complete the coursework and the Course 2 student project, you will have a valuable professional certification as a Certified Process Expert, not just a certificate of completion, and you will be awarded in many other training courses to follow.

NOTES

1 For each required exercise, the candidate will be challenged to answer "what immediate issue would be of the greatest benefit to the company if it were solved quickly and permanently."
2 QM does not imply that organizations must make the wholesale change to BMS but that the candidates define the most ripe and productive areas of the organization for improvement.

Appendix B: Terms and Definitions

Business Pathology
Identifying and removing foreseeable risk from within an organization.

Forensic
The application of scientific investigative methods and techniques to business operational health.

Foreseeable Risk
A danger that a reasonable person should anticipate as the result of their actions (legal definition).

Organizational Negligence
A legal term for the degree of neglect an organization exhibited in delivering their products or services to the stream of commerce.

Pathological Organization
A learning model for understanding that organizations are living organisms, each with its own personality and pathology.

Products Liability
A legal term defining the liability of any or all parties along the chain of manufacture of any product for damage caused by that product or service.

Quality
Fitness for use.

Quality as a Profit Center
A proprietary to evolve overhead quality systems into more robust systems that yield a return on investment.

Quality Management System (QMS)
A structured conformance system such as ISO 9001:2015.

Quality Professional
An individual who has two or more years of experience in quality systems and is skilled in the use of quality tools.

Quality System
Any of the various techniques evolved over the decades to control quality outcomes.

Risk
Exposure to danger, harm, or loss.

Risk Avoidance
The breakthrough process for avoiding risk instead of managing it.

Risk Management
Any of several schemas designed to manage risk at all levels; traditional risk management is typically prescriptive and reactive instead of strategic and proactive.

Situational Awareness
Perceiving the environment, comprehending the situation at hand, and projecting future status with the goal of removing human error.

Six Sigma
A set of management techniques to improve business process processes by reducing the probability that an error or defect will occur; it assumes a certain number of defects will always occur.

Standard of Care
The degree of care a reasonable person/organization exhibits to prevent harm or loss to another (legal term).

Strategic Thinking
An intentional and rational thought process that focuses on the analysis of critical factors and variables that will influence long-term success.

Appendix C: Strategic Thinking and Situational Awareness

The Strategic Thinkers' Compendium©

QUALITY MASTERS' CERTIFICATION PROGRAM©

Introduction

In years past, whenever I heard the term strategic thinking, I summarily patted myself on the back as "one of those" without a second thought. I suspect many of you do the same thing.

As my career as an expert witness in products liability and organizational negligence grew over the last couple of decades, I became obliged to find out what I was doing differently. I was amassing a 96% success rate in positive case settlements for my client lawyers. Some of my cases were settled after I produced an expert report. Some settled after I was deposed by opposing counsel. A handful went to trial, where I typically prevailed when I was called to testify.

What was going on? Was I that smarter than dozens of attorneys? Ego would like me to believe it true, but there must be more to it than an innate talent. There must also be other issues in play than the plaintiff or defendant having a rock-solid case. They seldom did.

As I started my self-assessment, I was drawn to articles on strategic thinking and began looking for answers. I was taken to study those who play high-stakes Texas Hold'em Poker. Their tournament plays on TV seemed to fit my synthesized definition of strategic thinking being an "intentional and rational thought process that focuses on the analysis of critical factors and variables that will influence long-term success." The players must have had some schooling in the nuances of this definition. Or so it would appear.

Since Texas Hold'em is a conscious exercise of lying and bluffing a player's way to victory, I was certainly not going to exemplify strategic thinking and situational awareness by using that game as a model. But, watching their play, I certainly could see their acute awareness of every nuance of each hand and the minutest of details in the body language of the other players. True winners used every sensory talent, combined with the mathematics of card-deck permutations, to create their strategy for long-term success. They must be hybrids of the four learning styles: Visual, Auditory, Kinesthetic, and Written Word, I postulated.

In my research, I saw subtleties such as smoking cigarettes and cigars have been being phased out in most competitive poker rooms. There appears to be an outright ban on them in tournament poker. Having smoke wafting through the air would certainly be a valid distraction to some and an advantage to others, especially kinesthetic learners. This is a microscopic detail of strategic thinking and situational awareness, but it can contribute to a complex equation leading to victory or defeat.

In my early legal cases, I became aware that making a case as an expert witness was an extension of conducting quality audits at manufacturing companies. Since most cases involved products that failed and caused great loss or harm, there is obviously a connection to operational integrity, competence, and process variability. Indeed, as I worked with the other causes-and-origins experts, failure modes quickly became obvious, but not as a rock-solid case for either side. As with poker, the difference in winning or losing a legal case had to do as much with the talent of the experts, lawyers, their effectiveness in reading body language, and how they conducted their theatrics.

The next thing to look at was my expert reports. Was I a spinner of tales with the talent to mesmerize readers at the level of J K Rawlings? Hardly. My training as an engineer taught me that all data must fit into rows and columns. There is no sleight of hand in a spreadsheet. I had obviously developed a talent for writing after publishing 12 books, but book sales were a minuscule amount of money to my bottom line. So, I would not call that career in and of itself successful.

Expert Reports

My first expert report was written about a case that involved a large aerospace manufacturer suing a small supplier of a material used in forging titanium parts for jet engines. The plaintiff only bought several hundreds of pounds of this material a year. But how, in this case, could critical destructive inclusions have made it all the way from the foundry to the customer without being detected? Something in the manufacturer's AS9001 QMS was breaking down.

I kept requesting additional production (documentation) from the plaintiff's company. I started following a thread of how the small provider had initially gotten on the approved supplier list. I obtained their history over a decade on the ASL. The supplier got on the approved list by filling out a desk audit. From that questionable beginning, I created a Visio diagram depicting each year that QA continued their approval.

There were attempts over the years to send an auditor and source inspector to the supplier's factory, but they never happened. The supplier would always respond with, "You do not buy enough product for us to host an audit." Their approval was continued for years by a series of obvious oversights that started with the initial approval.

The batch of products that contained the destructive inclusions was the result of a broken incandescent light bulb over the storage bin. It dropped tungsten chards into the mix. At the end of my investigation, it was revealed that the supplier made products for the cosmetics industry, not for aerospace. Yet this batch of products somehow made it through receiving inspection, where it was given a waiver for fluoroscopic inspection. Evidence showed that they did not perform this crucial testing step because they never had problems with this supplier over the preceding decade.

The strategic thinking lesson here is that the quality of the current shipment should never be assumed to be compliant based on previous shipments. For example, in the website history of the Florsheim shoe company, they conveniently forgot to include the time between 1984 and 2002 when they had the shoes made offshore. The loyal customers, including me, did not detect the degradation in quality for several years until it became apparent that these were not your father's Florsheim shoes. Those typically lasted for generations.

Next, the batch of material with the destructive inclusions made it through fabrication, assembly, testing, and shipment with the included flaw. Fortunately, the end users found the defects before the parts were installed on the aircraft. My expert report produced a PowerPoint presentation showing the nonconformances in the supplier approval timeline, the violations of AS9100, and their own procedures. I presented it in Federal Court, and the case was ruled in my client's favor.

Yes, my expert reports were a manifestation of strategic thinking and situational awareness. They were unlike any other expert reports presented in other cases. Opposing counsels' experts in quality typically attempted to transfer the blame away from the manufacturer. But, they were never successful in presenting incontrovertible evidence obtained through diligent process auditing. I discovered that I was validating the definition of strategic thinking as an "intentional and rational thought process that focuses on the analysis of critical factors and variables that will influence long-term success."

INTERMISSION

If this introduction appears lengthy and convoluted, it is because becoming an expert in strategic thinking and situational awareness is a protracted and intricate process. Please read on. It will help you perform your own personal self-assessment and your career path going forward.

DEPOSITIONS

The next area of study was my performance in depositions. Opposing counsel would have retained another quality expert to, hopefully, contradict my expert report. Counsel will have prepared questions about my report and from the other experts, as well as a barrage of issues about my credentials and integrity.

I can say, with all candor, that opposing counsel seldom gained any yardage in disproving my opinions. Nor were my credentials ever successfully challenged. I was subject to depositions that lasted a full day (as opposed to the two hours scheduled). The protracted examinations just made opposing counsel increasingly disposed to keep grinding over the same issues, expecting me to change my testimony. To this day, I look forward to being deposed. It is an excellent test ground for strategic thinking and situational awareness.

The lessons learned were that I, again, prepared for depositions by looking at all variables that could contribute to long-term success. By this time, I was being handsomely compensated for my time, but the combined efforts of the legal teams most often shortened the litigation process and created compelling arguments that led to favorable outcomes.

OUTCOMES

Over time, I reverse-engineered my expert witness experiences into my career as a quality professional, QMS implementer, and business process facilitator. After witnessing pictures of many dead and disfigured people, I concluded that the foundational teachings of our profession were *fundamentally flawed*. We were in the business of minimizing process variability and defects, but we were not in the business of avoiding them and driving critical outgoing defects to zero. This was a key motivation in creating the QM training program.

In this compendium, I will present tools of strategic thinking and situational awareness to help you evolve your innate skills. In the QM certification courses, you will learn how to implement business process excellence and risk avoidance. You may also find yourself looking for a career as an expert witness when you complete Course 5. This role can be both professionally and financially rewarding.

LEXICON

> **Strategic Thinking**—An intentional and rational thought process that focuses on the analysis of critical factors and variables that will influence long-term success
>
> **Situational Awareness**—Being aware of the environment, comprehending the situation at hand, and projecting future status with the goal of removing human error
>
> **Business Management System (BMS)**—An enterprise-wide system of risk avoidance and business process excellence

PROLOGUE

I became aware of situational awareness as I aged and grew more vigilant. For the last 25 years, I have driven a winding mountain highway that rises 2,200 feet in elevation of seven miles. The road is the only thoroughfare from Reno to Historic Virginia City, NV. It is driven by thousands of tourists year-round. It provides a mandatory survival study in situational awareness. Not only do I, who use the road daily, need to have mastered the twists and turns, but I must be acutely aware of tourists who are unskilled in mountain driving. Some drive at 20 MPH when the posted speed limit is 45 MPH. Some get so petrified that they stop on the roadway. The Sheriff is known to have had to rescue some of them. Snow and ice exacerbate the danger for the visitors and the rest of us.

I have never had even a narrow escape on that highway. My eyes are continually scanning every nuance of the road and prevailing traffic. I've learned to "look around" the next turn in the road. I make it a point to be aware of every potential indicator of foreseeable risk. When traffic permits, I judge my situational awareness as being able to negotiate the seven miles averaging the speed limit. At age 79, I continue to navigate the road with precision and skill. If I ever detect any change in my acuity, I will hang my keys up.

On the city streets and freeways, I am blessed with exceptional peripheral vision. I am continually aware of what is happening within my forward and rear-viewing field. In 60+ years, I have had only two minor collisions in inclement weather, which were deemed the other driver's fault. As a strategic thinker, I have synthesized my personal driving into key elements of situational awareness training and included them in this compendium and in the QM courses.

QMS TO BMS

The first application of strategic thinking and situational awareness is within the fundamental paradigm shift included in the QM training. That is, we need to begin thinking strategically about the transformation of our Quality Management System (QMS) into an integrated Business Management Systems (BMS). We were inculcated with ISO 9001 as the framework for a QMS as soon as we were introduced to the Standard. It has become part of our cultural foundation. It is now time for that to change.

While the Standard holds itself as the requirements document for a certified QMS, it is grossly incomplete as a comprehensive tool for running a business enterprise. While it repeatedly discusses continuous improvement, it focuses on QMS conformity. Most implementations are centered on attaining or renewing a certificate. This is an example of myopic thinking and a lack of situational awareness and strategic thinking. Often, the QMS takes on a life of its own, often with excessive documentation and focus on audits leading to achieving compliance.

To become champions of business process excellence and risk avoidance, we must examine our methods and beliefs, much as I did in conducting expert witness assignments. One of the most challenging strategic thinking exercises is looking beyond our paradigms and assessing how we can transform what we know and the tools we use to implement enterprise-wide realization. *We must put on our strategic thinking and situational awareness hats every time we practice our trades.*

In QM training, we emphasize a rapid transition from traditional quality audits to Forensic investigations. In the Forensic Investigations Handbook, I devote an extensive amount of the teachings to highlighting the shortcomings of traditional QMS auditing. We then move on to the fundamentals of Forensic Investigations. This evolution to strategic thinking requires that we see internal audits as the parochial exercises that they are. We are not implying that they are ineffective for QMS compliance; I am stating that they are inadequate for achieving business process excellence and risk avoidance.

THE INITIAL COMPLIANCE INVESTIGATION

The first exercise in conducting an enterprise-wide investigation and assessment is not a trivial exercise. It cannot be accomplished in a one-day or one-week audit. You, as a champion, must make your defensible case to senior management that what you have learned in your QM training is the launching pad for the future success of the business. Your investigation and proposed BMS implementation will have metrics tied to the KPI of the organization. Some forward-thinking practitioners have already begun creating KPIs at the process and department levels. Their next challenge is classifying them into business KPIs.

It is highly recommended that you impanel a Process Improvement Team (PIT) as soon as you are ready to begin the transformation. You can use the template provided to create your own tool or obtain the full evaluation tool from Productivity Resources. Including each assessment point and adding your own unique organizational norms is a key to success. Addressing organizations as unique living organisms has always been missing from the ISO 9001 cookie-cutter structure.

The PIT must be trained and deemed competent in the tools of the Forensic Investigation Handbook. Investigative assignments must be made by those familiar with the areas to be assessed, along with input from upstream and downstream individuals representing their organizations. Acknowledging that you are moving outside the confines of the QMS, these investigations could take weeks of effort from individuals who already have full-time jobs.

For consultants, this same approach must be conducted at the onset of an assignment. You will have significant success by mapping out the potential rewards for the senior managers before launching initial compliance investigations. In fact, I typically insist on an off-site retreat for both organizational and consulting assignments. The management team will then take ownership of the project, allow PIT members to do the assigned work, and hold them accountable for their timelines and work outputs.

This is also the time for you, as the champion, to go back and revisit the definitions of strategic thinking and situational awareness and exactly how you are going to train your team to use these tools. This is an extremely beneficial exercise for producing a comprehensive investigation and for the members using these tools in their daily lives. As presented in my examples of driving proficiency, they can be tools for improving quality of life as well as for business excellence.

THE IMPLEMENTATION PLAN

The implementation plan must have at least two components. The first is how to implement business process excellence and risk avoidance into each organizational element. This must include the findings of the initial investigation and plans for whatever remedial or evolutionary findings were concluded. The second part is a map of the interrelationships of each organizational element with the processes they interface with and those required in the overall business process flow. This must be a living plan that continually evolves with the organization and with the ongoing success of the BMS.

The most effective tool for long-term success is to have the PIT become a perpetual committee with revolving membership. Its duties would include the overall task of continual improvement plus conducting forensic investigations when needed.

STRATEGIC THINKING AND SITUATIONAL AWARENESS IN PROCESS COMPLIANCE

Throughout this courseware, I repeatedly stress the theme of moving past the goal of conformance to the Standard to process excellence and risk avoidance. This does not imply that organizations do not desire to remain compliant with the Standard. The fundamental truism is that if you follow the BMS 9001:2023 Guidance Document, *you will innately comply with ISO 9001:2015.* I have subsumed the Standard in the QM Training Program. There will no longer need to be a separate conformance exercise for internal auditing and third-party auditing. When I implemented the Business Management Interactive System at Dell Computer during their 2000 transition, their registrar cited their BMS system as best in class.

FORENSIC INVESTIGATIONS

I continually stress shifting paradigms of how I have traditionally approached compliance audits into forensic investigations. The definition of forensic investigations is utilizing scientific investigative techniques in assessing process conformity, needed corrective action, and continual improvement. This evolution is critical to achieving business process excellence and risk avoidance.

In applying strategic thinking to our classic methodologies for audits, FMEA, and Root-Cause-Analyses, I must step up to a 50,000-foot perspective of what we were trying to accomplish with each of these tools.

Audits are typically an analysis of processes to determine compliance with a standard. Strategic thinking opens horizons of opportunity by examining the competency of the individuals being interviewed. Looking at it introspectively, the classic audit exercise is an ideal time to determine if the operator is indeed competent compared to their Roles and Responsibilities. I doubt very many seize this opportunity.

Another strategic thinking view of internal audits is that everyone knows exactly what the scope of the audit is, and they are typically well-rehearsed in the correct responses. We are not suggesting a return to punitive surprise audits, but forensic investigations provide a clear scope of their purpose. Instead, we leave comprehension of the processes involved up to those interviewed to describe in their own words. In taking this investigative approach, all facets of competence, awareness, training, and demonstration of the soundness and understanding of the processes are examined. It encourages situational awareness by observing the nuances of the process and the operators to maximize the effectiveness of the investigation. The outcome is more than a mark on a checklist; it is a more complete evaluation of compliance *and* effectiveness.

FORENSIC INVESTIGATIONS 2

Situational awareness in conformance investigations manifests in a much more in-depth than traditional analysis, such as we might conduct of the purchasing process. Traditional audits would examine a handful of purchase orders, looking for process compliance. Forensic investigations are much more comprehensive. YES, there must be a fundamental evolution from internal audits to forensic investigations. YES, these investigations are conducted much differently than internal audits. YES, they require moving past the auditing techniques I described in years past in many of my books. YES, these investigations must be designed to match your exact organizational culture. YES, there is a measurable return on investment to be realized from forensic investigations as opposed to the overhead exercise of internal audits.

Forensic investigations are conducted to determine if a particular process, or set of processes, is being conducted as the relevant procedures and documentation prescribe. They are like the current PFMEA's but conducted from a forensic perspective. That is, the tools of FMEA may not be appropriate for all Forensic investigations. Since we are implementing new paradigms of process excellence and risk avoidance, there may be alternative investigative and tabulation techniques that are more effective. Also, we are conducting these investigations enterprise-wide, so QMS tools may not be appropriate for other departments and activities. As we are implementing cultural shifts in the organization, we must redefine our tools.

Also, PFMEA's are often conducted by the people who are directly involved in the processes. These investigations may become incestuous. The investigators may not be able to look beyond their paradigms to see the root causes. Forensic investigation teams are best composed of multidisciplinary groups that understand the processes but are not all involved in their implementations.

There are many guidance documents suggesting the most effective ways to conduct root-cause analysis (RCA). Most are helpful in creating the discipline for those involved to ask "the five-why's" until they have discovered the true root cause. Putting on your new strategic thinking and situational awareness hats, there is a new universe of discovery possible in performing RCA's.

There is, of course, the fundamental question of how you know that you have found the root cause. Are more than five whys needed? The accepted maxim is when there is no reasonable return on investment in probing further. Strategic thinking suggests that you look beyond the immediate issue to how surrounding influences contribute to the problem at hand.

One manufacturing company I worked with was having issues with assemblers placing the wrong value resistors in through-hole circuit boards. The RCA investigation kept circling around to the assumption that the assemblers were not trained fully on reading the bands of color codes that defined the ohmic value of the resistors. Moving beyond the obvious answer that increased training would solve the problem, further input from the inventory and service personnel suggested that they use color-coded markers, along with the values, on the bags containing the resistors. Taking the guesswork out of what the color bands should look like solved the problem.

Unfortunately, that did not solve the problem of the service techs breaking into the stockroom on the weekends, harvesting service parts without them being signed out of the inventory control system. That RCA solution came from the custodian, who suggested building them a separate and secure service inventory of common parts outside the stockroom. These components were recategorized as consumables, and the inventory integrity returned to 98%. I also had to secure the ceiling tiles where they were crawling through to gain access to the stockroom.

There is an elegant example of strategic thinking from another company that manufactured desktop computers. As each one was built, they were moved into a holding area on rolling carts. By the end of a shift, there would be a sea of white computers needing to be sorted by next-day shipment and second-day shipment priorities. The only place this information was available was on the traveler attached to each one. The manufacturing engineers tried in vain to find a solution to sort them out before going to shipping. One of the assemblers took it upon himself to go to a hobby store and purchase a gross of wooden dowel rods. He painted one end blue and the other red. They fit conveniently in holes in the carts. The red ones, indicating the next day, went on the left side, and the blue ones went on the right side of the carts. At the end of each shift, it took only a few minutes to separate them into the correct shipping order.

STRATEGIC THINKING AND SITUATIONAL AWARENESS IN PROCESS IMPROVEMENT

Improvement investigations are the most bountiful for achieving breakthrough outcomes in process excellence. If you impanel cross-functional PITs for each investigation, you will certainly discover solutions that were not apparent to the usual suspects in your typical continual improvement efforts. The themes for these efforts are typically the result of audit outcomes, management reviews, and suggestions from concerned team members. In any event, the input is often that a process currently flows using steps A, B, and C, but it would be much more efficient if I added step D.

For this section, I will tackle the major improvement project of evolving a QMS into a BMS. You will need these practical suggestions to begin this formidable task in your organization or for a consulting client. The archetype will be the example of developing the Business Management Interactive System at Dell Computer during their ISO 9001:2000 transition. I have published some of the results in other QM Courses but will need to delve more into the nuts and bolts of how this transition was realized. The process improvement initiative is just as valid as a blueprint today as it was 20 years ago.

CHECKLIST FOR IMPLEMENTING THE STRATEGIC THINKER'S BMS MODEL

The Imperatives

- All business activities must be viewed through the lens of foreseeable risk.
- Each process has a finite and predictable number of potential variables that must be defined and assessed for improvement and foreseeable risk.
- Humans have the capacity to create, hide, and exacerbate potential defects.
- Familiarity prevents us from seeing all risks within our products and services.
- It is possible to avoid risk rather than manage it.
- Traditional quality management tools are designed to minimize nonconformities, not to eliminate them.

The Foundational Tenets

1. **Leadership**
   a. Vision, Mission, and Values
   b. Effective Communication
   c. Training and Mentoring
   d. Stewardship
   e. An Ethical Compass
   f. Fiscal Stability
   g. Succession Planning
   h. Risk Avoidance Mandate
2. **People**
   a. Awareness and Training
   b. Stakeholders and Shareholders
   c. Accountability and Ownership
   d. Rewards for Excellence
   e. Risk for Underperformance
   f. Outcome Focused
   g. Self-Directed
3. **Processes**
   a. Designed and Defined
   b. Planned and Executed
   c. Supply Chain Excellence
   d. Defect-Free Product Realization
   e. Verification and Validation
   f. Continual Improvement
   g. Data/Software/Firmware Integrity
   h. Risk Free Operations
4. **Metrics**
   a. Return on Investment
   b. Key Performance Indicators
   c. Detailed Process Metrics
   d. Voice of the Customer
   e. Noncompliance Data
   f. Time Utilization
   g. Improvement Data
   h. Risk Indices
5. **Products and Services**
   a. Benchmark of Quality and Reliability
   b. Fitness for Use
   c. Intrinsic Value
   d. Regulatory Compliance
   e. Inherent Safety
   f. Defect Free

6. **Outcomes**
   a. Defect-Free Products
   b. Error-Free Services
   c. Value Exceeds Cost
   d. Liability Free
   e. Standard of Comparison
   f. Do No Harm
7. **Customers**
   a. Partners
   b. Demand Value and Defect Free Products
   c. Receive Exceptional Service
   d. Give Continual Feedback

STRATEGIC THINKING AND SITUATIONAL AWARENESS IN PHASING OUT RISK ASSESSMENT

There are scores of approaches to what is commonly called risk management. There are even more tools in the public domain for conducting risk management assessments. There are organizations of consultants who will be happy to take your money to conduct their versions of risk management. As I have stated throughout the QM courseware, traditional risk management lacks the fundamental mandate to avoid risk. ISO 31000 has added even more compliance fodder that obfuscates risk avoidance.

Locating foreseeable risk in businesses is a complex process of strategic thinking and situational awareness. It is not always effectively discovered through checklists and matrices. While I encourage you to find the tools that will assist you best in risk assessment, I propose that you think outside the box when conducting these investigations. In fact, my mantra from my years at NASA is to "think outside the globe."

ADOPTING THE RISK AVOIDANCE PARADIGM

For organizations to be successful in risk avoidance, there must be a top-down paradigm shift in the vision, mission, and value system of the company. Whether your company culture has morphed organically over time or is the result of enlightened strategic planning, the concept of risk avoidance is seldom part of the company mission.

The fault is not one of pragmatic process development or selecting one quality management approach over another; it is embedded in the fundamental paradigms of organizational leadership. Risk avoidance is not taught in B school or in management seminars. It is not part of business or quality management training curricula.

It is a new paradigm that states that defects are avoidable, not inevitable. If your value system is based on the inevitability of some level of nonconformities being acceptable, you will always have nonconformities.

Unfortunately, since ISO 9001:2015 added risk-based thinking to our lexicon, quality professionals have attempted to create compliance parameters for risk-based thinking rather than taking a 10,000-foot level assessment of its meaning and implications. The following steps are not found in your father's quality management system.

1. Revisit your Vision, Mission, and Values
   a. Senior management must become fully aware and enlightened on the tenets of risk avoidance and foreseeable risk.
   b. If you are to be successful, the vision, mission, and values must be revised to include a mandate for risk avoidance.
   c. The new mandates must be communicated at all levels and incorporated as a culture, not a requirement.
2. Reassess your quality management system
   a. While ISO 9001:2015-based QMS addresses business success, implementation is typically founded in the traditional steps of continually improving processes to achieve more favorable outcomes and metrics.
   b. Determine where your organization falls in Table 16.1 and determine for which level you are initially striving.
   c. The QMS reassessment process must be one of changing the focus to quality as a profit center, stupid proofing processes, and products, metrics driving key business indicators, avoiding defects reaching the customer, and achieving perfect customer report cards.
   d. Process control must include inculcation of the vision, mission, and values.
   e. Each tenet of your QMS must be restated from risk management to risk avoidance guidance.
   f. There are proven tools that have been developed for risk avoidance.
3. Implement the Risk Avoidance business model
   a. Conduct presentations and workshops to teach Foreseeable Risk.
   b. Assess your compensation program. Consider a risk and reward program that has incentives for defect-free performance and penalties for creating nonconformities.
   c. Recast your version of PDCA to replace the "check" function with an "avoidance" function.
   d. Assess the control limits of your metrics to strive for zero nonconformities, not an acceptable number.
   e. Utilize every defect report, customer input, and corrective action as gifts for your new goal of delivering defect-free products and services.

The Tools of Foreseeable Risk

Assessing foreseeable risk is the next level in quality management. It is not reinventing your business processes or adding more overhead to ensure conformance.

Before delving into the tools, take a moment to study Table 14.1. It is included to help with a paradigm shift from the requirements of ISO 9001 to a business-level view of risk avoidance.

1. Assess and revise the steps of each process in your organization. Include two mandatory questions for the process operator to answer before moving the product or service to the next process step.
   a. Am I certain that my work is defect-free before passing it to the next process step?
   b. Am I solely accountable to the next process operator for my work product?
2. Add a step to audit and management review meetings. Ask the questions
   a. How can the owner of the next process use my work output badly or inappropriately?
   b. How can the work output of each process affect a customer?
3. Add a step to your FMEA processes or empanel a risk avoidance committee. Include individuals who have no formal involvement in the processes but are dedicated to the success of the company. Add these steps:
   a. During DFMEA's's and PFMEA's add a round of questioning about how a customer could use the product or service badly, stupidly or for the wrong purpose.
   b. Review customer complaints, warranty reports, and repairs with a deliberate focus on what has been done to avoid each issue. Customer stupidity will not be a relevant factor in a lawsuit if you have not anticipated potential misuse.
4. Continually model the culture that a defect will never leave my workstation; therefore, a defective product or service will never reach a customer.

STRATEGIC THINKING AND SITUATIONAL AWARENESS IN FORENSIC INVESTIGATIONS

Success as a consulting or testifying expert is 50% forensic science and 50% strategic thinking and situational awareness. Whether you are representing your company as an internal expert or practicing in litigation, the ability to see beyond the available data and maze of contradictory evidence is critical to reaching a defensible opinion.

Situational awareness places you inside the catastrophe or disaster as an observer free from harm and bias. It positions you in the lookout's Crow's Nest, with a clear 360-degree view of all events, even though the boat is continually pitching, yawing, and rolling. You can see through the mist and fog hazards from port and starboard and from fore and aft. Without fear of drowning, you can call all events exactly as you see them.

You can question every decision in navigation and operation with the certainty of a Monday-Morning Quarterback. Utilizing strategic thinking, you can examine each event with clarity and precision if your investigative talent allows precision assessments of the environment surrounding the event.

Some years ago, an attorney retained me to conduct the principal investigation of an incident at a senior living community. The lawyer knew I had experience with AC power distribution because I had been involved in building two new buildings for companies that I worked for. I obtained firsthand knowledge of power distribution, HVAC, and fire suppression as part of the design and implementation teams.

I have also been a licensed ham radio operator since I was age 13. Over the years, I have had extensive experience with electronics, high-power equipment, and safety issues. *This information is critical for you to understand how your work and life experiences can become part of your resume.* My background gave me a working knowledge of the incident, although I was not a licensed electrician or having worked in that field. We do not have to be experts in a field to investigate and render opinions on process failures and human error. We need to be situationally aware and able to conduct precision forensic investigations.

The incident involved an electrician who was shocked working on a 4,000-volt distribution panel. He sustained severe burns and injuries, rendering him permanently disabled. When I arrived, I began examining the evidence that remained after the accident. I found a "blown" 400-amp fuse from the power panel. Examining the power panel and fuse, I found evidence of electrical arcing that left clear carbon trails from what must have transpired. I found wire leads from a hand-held test meter. One lead was charred on the bottom of the cabinet, while the other was draped over the electrical bus bars.

Most electricians carry these meters for use in diagnosing electrical problems. I have one in my toolbox that is exactly like the one in this incident. They are designed to be used in circuits that are less than 600 Volts. It had no place in a 4,000 Volt distribution panel. From pictures taken by the staff after the incident and their testimony, the root cause of the incident became obvious. For some reason, the electrician defeated the safety lock on the power panel and used his meter to determine if a fuse was blown. His critical error was that the high voltage was still present even when the power panel was in the disengaged position. The high-voltage carbon arcs painted a picture of how he tried to troubleshoot the fuse. The burns on his hands and body collaborated my assessment. I will never know why an experienced electrician violated every safety procedure and common sense move in electrocuting himself. His situational awareness was grossly lacking on that fateful day.

The cause of the incident was clear. The diagnosis was validated by the fire department and another company that specializes in high-voltage power distribution. The case was open and shut. On the last day before the statute of limitations ran out, his family filed suit against the company that owned the retirement community. Five years later, the litigation is still ongoing, and my incident report was rewritten as an expert report.

QM Course 4 contains seven case studies from my files. You should revisit them from the perspective of strategic thinking and situational awareness.

STRATEGIC THINKING AND SITUATIONAL AWARENESS AND REAL-LIFE FORESEEABLE RISK

The following are some case studies from the public domain that highlight accidents and disasters that were completely preventable if they had been students of risk avoidance, strategic thinking and situational awareness.

HONOLULU, HI (AP)—*Federal investigators blamed a deadly 2019 helicopter crash in Hawaii on the pilot's decision to keep flying into worsening weather, and in a report Tuesday, they accused regulators of lax oversight of air tours that are popular among tourists to the islands.*

The National Transportation Safety Board said that the Federal Aviation Administration had delayed installing aviation weather cameras that might have alerted the pilot to the fog-shrouded conditions in a mountainous region on the island of Kauai. The board also said the FAA failed to do enough to ensure that your pilots in Hawaii are trained in handling bad weather.

(BRITANICA)—*The worst nuclear accident in U.S. history began at 4:00 a.m. on March 28, 1979, when an automatically operated valve in Three Mile Island's Unit 2 reactor mistakenly closed, shutting off the water supply to the main feedwater system (the system that transfers heat from the water circulating in the reactor core). This caused the reactor core to shut down automatically, but a series of equipment and instrument malfunctions, human errors in operating procedures, and mistaken decisions in the ensuing hours led to a serious loss of water coolant from the reactor core.*

GRANTSVILLE, UT (AP)—*A woman died in an industrial accident at a mattress manufacturing plant in Utah, according to police. Police were told the woman was standing by a machine and got "pulled in" Thursday afternoon at the Purple Mattress factory in Grantsville, Sgt. Jeff Watson told KUTV.*

CLAIRTON, PA (SUNNEWS)—*A Pennsylvania manufacturer's failure to provide guarding on a brick-crushing machine ended in a 53-year-old worker suffering an arm amputation while operating the machine.*

A subsequent federal investigation at TYK America Inc.'s Clairton refractory products manufacturing facility found the company previously identified a deficiency with the machine's guarding but kept the machine operational and did not repair the deficiency prior to the amputation incident.

(Mpora)—*In 1997, Police in Reston, Virginia, issued a statement saying they had found the body of 22-year-old Eric Barcia, who had died attempting to bungee jump off a 70-foot bridge. Eschewing commercial bungee operations, Eric had taken matters into his own hands and tied several bungee cords together. He strapped himself securely, tied the other end to the bridge, and jumped, confident in the knowledge that he'd carefully measured out the bungee's total length—just under 70 feet. Of course, what Eric had forgotten was that bungee cords stretch . . .*

Appendix D: Forensic Investigation Handbook

CONDUCTING FORENSIC INVESTIGATIONS

THE NEW GOLD STANDARD IN PROCESS AUDITING

Conducting Forensic Investigations

The New Gold Standard in Process Auditing

Author – Tom Taormina, CMC, CMQ/OE, QM

Contributing Editor - Steven Garner, QM

Quality Masters' Certification Program

Exemplar Global Certified Training Provider

Revision 0

Contents

Introduction

Forensic Investigations are more robust and beneficial than traditional management systems and process audits. They not only discover the degree of process compliance, but they go on to identify opportunities for business process excellence and risk avoidance.

Underlying definitions used in this document include business, quality and legal concepts including:

Business Management System (BMS) – An Enterprise-wide system of controlling and improving all business processes under one coherent infrastructure.

Business Management Professional – An individual who has completed certificate courses one through four of the Quality Masters' Certification Program[1].

Business Pathology - The study and diagnoses of business process systems that go beyond traditional root cause analysis and discover any foreseeable risk and potential liability issues.

Business Process Excellence: Enterprise-wide optimization of business processes and their interrelationships designed to improve key organizational goals.

Business Process Management - Every activity is a process. Every process needs a defined structure. Every process must be integrated into a logical and integrated Business Management System.

Certificate - Receiving written acknowledgement from an authoritative body that an individual has completed required coursework and successfully passed all exams within the scope of any given course.

Champion of Business Success - Quality professionals who have successfully completed all the coursework of The Quality Masters' Certification Program and are facilitators that enhance the strategic goals of an organization.

Critical Defect - As defined by your organization and culture, a defect that can never reach a customer.

Diplomate – An individual who receives a diploma from the Quality Masters' Program. It connotes successful completion of a course, with validation by subject matter experts.

Duty of Care - The ethical responsibility of a person or organization to avoid behaviors or omissions that could reasonably be foreseen to cause harm or loss to others (legal term[2]).

Forensic Investigation - An enhanced version of proactive management system auditing. It includes evaluation of business process effectiveness and utilizes the tenets of foreseeable risk to uncover opportunities for risk avoidance. It also can be used as a powerful tool for competence evaluation.

Forensic: The application of scientific investigative methods and techniques for business operational health.

Business Pathology – Identifying and removing foreseeable risk from within an organization.

Foreseeable Risk: A danger that a reasonable person should anticipate as the result of their actions (legal term).

Quality Master Certification Program - The five-tiered program that progressively trains and certifies individuals to become the consummate Master of business process excellence and risk avoidance.

Quality Management Practitioner (QMP) – An individual with more than two years of experience in the quality disciplines and has completed Course 1.

---

[1] Qproductivity.com - Quality Masters' (QM) Training and Certification Program

[2] Legal terms are included because they are the accepted definitions by which an organization and individuals will be judged in a court of law. Risk Avoidance is key to preventing litigation.

Process Management Expert (PME) - A comprehensive credential for quality professionals that certifies an individual is qualified to operate at an advanced level in QMS implementation, auditing, and improvement. An individual who has successfully completed Courses 1- 2.

Business Process Excellence Master (BPXM) – An individual who has completed Courses 1 - 3.

Business Risk Avoidance Master (BRAM)- An individual who has successfully completed Courses 1- 4.

Organizational Negligence - A legal term that describes the degree of neglect an organization exhibited in delivering their products or services to the stream of commerce.

Pathological Organization© - A learning model for understanding that organizations are living organisms, each with its own personality and innate pathology.

Products Liability - A legal term defining the liability of any or all parties along the chain of manufacture of any product for damage caused by that product or service.

Quality as a Profit Center[1] - The proprietary methodology developed by QProductivity to evolve overhead quality systems into more robust systems that yield a return on investment.

Quality Management System (QMS) - A structured conformance system such as ISO 9001:2015.

Quality System - Any of the various techniques evolved over the decades to control quality outcomes.

Risk - Exposure to danger, harm, or loss (legal term).

Risk Avoidance© - The breakthrough process for avoiding risk instead of managing it.

Risk Management - Any of several schemas that have been designed to manage risk at all levels. Traditional risk management is typically prescriptive and reactive instead of strategic and proactive.

Situational Awareness – Being aware of the environment, comprehending the situation at hand and projecting future status with the goal of removing human error.

Six Sigma - A set of management techniques intended to improve business processes by reducing the probability that an error or defect will occur. It assumes a certain number of defects will always occur.

Standard of Care - The degree of care a reasonable person/organization exhibits to prevent harm or loss to another (legal term).

Strategic Thinking - An intentional and rational thought process that focuses on the analysis of critical factors and variables that will influence long-term success.

---

[1] Codified and proven at Dell Computer as BMIS- The Business Management Interactive System

Background

Forensic Investigations are a non-traditional approach for evaluating process effectiveness and determining risk within an organization. It is an amalgamation of tools and techniques that a company can use to enhance overall business excellence to help ensure that quality is at the forefront of all internal processes.

This course was derived from the author's 50+ years as a quality control engineer, consultant, and trainer having worked with more than 700 companies. It is also the product of nearly two decades as a consulting and testifying expert witness in products liability and organizational negligence.

Used as a business process excellence tool, application of forensic investigations within an organization can:

- Create processes that are virtually risk free
- Put metrics in place to track potential risks within each internal business process
- Achieve the goal of zero outgoing critical defects

As a risk avoidance tool, forensic investigations:

- Determine potential risks
- Assess all foreseeable risk factors
- Establish the tools of risk avoidance in all business processes

As a tool for determining appropriate or negligent standard of care, Forensic Investigations:

- Establish the appropriate and required duty of care
- Compare actions of an incident to the duty of care
- Build a legally compelling case for appropriate or negligent standard of care

Attributes

Since you are taking this course, you have already obtained certifications in quality deployment, quality management and process excellence. You have proven that you have the experiential background to observe, assess and make recommendations for process improvement and risk avoidance.

Over your career, you have amassed practical knowledge of a variety of technologies and activities that are integral to overall business operations. The depth and breadth of your experiences determine how valuable you are to an organization or to consulting and auditing clients.

The attributes that make you a successful Forensic Investigator include:

- The persistence and concentration necessary to collect all the facts of a given topic without jumping to conclusions prematurely
- The ability to assemble the facts and data into structured collections of viable evidence
- The ability to create preliminary observations of probable cause from your research
- The ability to collaborate with others to form a complete incident scenario and ask necessary questions about behaviors or omissions
- Acumen in turning the evidence into hypotheses
- Testing the hypotheses until a single logical scenario emerges
- Creating a written report that turns hypotheses into defensible conclusions and opinions
- Being able to authentically present your case verbally and in writing
- Being able to debate and/or mediate differing views and change your conclusions should a differing opinion prove to be more applicable
- Creating a final report that proves or disproves appropriate standard of care, as applicable

Finally, a Forensic Investigator MUST:

- Not become engaged in the technical minutia of the event. You must NOT be influenced by emotional involvement in the science or technology
- Be accomplished in strategic thinking
- Be a student of situational awareness
- Have the skills to prove what the evidence presents
- Have the fortitude to be disproven when the facts present otherwise
- Have the wisdom to know the difference

Traditional Process-Based QMS Audits

A common definition for the globally accepted scheme of periodic compliance audits is a systematic, independent, and documented process for obtaining objective evidence and evaluating it objectively to determine the extent to which the audit criteria are fulfilled. In the case of an audit of an ISO 9001:2015 QMS, the audit criteria are the Standard, internal operational procedures and subordinate QMS documentation.

As defined, classic QMS internal auditing is restricted in scope to the QMS and not inclusive of the overall business operations. Third-party QMS auditing is even more prescriptive since the auditing agency typically provides a written scope of the audit and must stay within that scope.

The output of these audits is most often simply an audit report. This report typically cites what was observed compared to the governing procedures. The audit checklist used to create the audit report is often filed as background information and not included in the report. Within the report are statements regarding the conformance of a process compared to its objectives. Unless a mature set of measurable metrics for key internal processes (e.g., Engineering, Supplier Control, Purchasing, etc.) have been deployed, these statements are typically more opinion than forensic evidence.

The auditor is then free to make comments on the effectiveness of the process against their own training and experiences. The auditor can add any observations for corrective action or improvement opportunities. These are typically experiential and not compliance oriented.

Finally, the audit report opines on whether the QMS is compliant with the Standard as a whole, or whichever Clauses are being audited. This conclusion is often mitigated with successful resolution of minor or major findings. The response to the findings should substantiate the conclusion of compliance or noncompliance. It is rare that a finding goes unresolved or that a QMS is disapproved.

Traditional audits only identify nonconformities to the requirements, not how they affect the overall performance. Forensic Investigations ask the question "Are the processes that are in place effective for their stated purpose?" This new approach to process evaluation evolved over two decades of applying quality auditing techniques to conducting forensic investigations in civil litigation. Essentially, the "five-why's" of root cause analysis are utilized during the interview process, not researched days or weeks later. There is also more rigor used in examining records presented supporting conformance AND effectiveness. The investigator asks for specific evidence, not that which was prepared and known-compliant in advance. Forensic investigations are the foundation for business process excellence and risk avoidance.

Typical QMS Audit Checklist

Figure 1 is a typical checklist for conducting ISO 9001:2015 audits[4]. It is annotated with numbers that correlate to questions about what is and should be included in "Audit Evidence." This narrative moves beyond the typical inclusions of evidentiary documents, procedures, work instructions, job descriptions and the like. This creates the first awareness of the differences between traditional audits and forensic investigations.

ISO 9001:2015 internal audit checklist for manufacturing companies

No.	ISO 9001 Clause	Audit Question	Compliant (Yes/No)	Audit Evidence
1	6.1	Has the company determined the risks associated with its manufacturing process?		1
2	6.1	Is there any evidence of the effectiveness of the actions taken to address the risks in the production process?		2
3	6.1	Does the company identify and implement opportunities in the production process?		3
4	7.1	If there is a maintenance procedure for production equipment and tools, has it been followed?		4
5	7.1	Are there any records for the maintenance of the production equipment and production tools?		5
6	7.1	Are gauge calibrations up to date?		6
7	7.1	Are the work environment and equipment appropriate for the job, safe, and clean?		7
8	7.2	Has the worker been appropriately trained on the job conducted? Are there records for this training?		8
9	7.2	Are suitable methods used to verify training effectiveness for the employees involved in the manufacturing process?		9
10	7.5	Is the necessary documentation available and suitable for its use in production? This includes work instructions, job orders, checklists, visual cards, etc.		10
11	7.5	Have workers received the most recent version of that documentation (work instructions, job orders, checklists, visual cards, etc.)?		11

Figure 1 – Typical ISO 9001 Audit Checklist

---

[4] From Advisera ISO 9001 Academy Free Downloads. Tom Taormina is a Certified Lead Auditor and Lead Implementer to ISO 9001:2015 by Advisera and is a contributing author to the Advisera Blog.

The following are challenges to the efficacy of traditional ISO 9001 audit checklists from Figure 1:

1 How can an auditor evaluate a global question about how the company may or may not have determined risks associated with its manufacturing or services process? This would involve an enterprise-wide risk assessment by trained examiners.

2 Again, this requirement for assessment of evidence risk evaluation is outside the experiential background and credentials of the typical ISO 9001 auditor.

3 What empirical evidence can an auditor evaluate to answer whether the company has identified and implemented opportunities? What is the definition of "opportunities?"

4 Determining if there is a "maintenance procedure" will typically be machine specific. Finding a sample should be easy. Determining if the procedure is being followed would require step-by-step auditing of the procedure as it is executed. As with #2 above, this is typically outside the expertise of an ISO 9001 auditor (also see #10 below).

5 The auditee should be prepared to offer any maintenance records. What does the existence of these records prove about the effectiveness of the process?

6 A sampling of devices to determine if gauge calibrations are up to date, is a missed opportunity. Less than 100% of active tool and equipment calibrations being current means the process is broken.

7 What empirical evidence can an auditor evaluate to determine if the work environment and equipment are appropriate? This can only be done at the end of a 100% audit.

8 Training records should be available for each process operator and owner. Determining the appropriateness of the training can only be determined by historical records of performance, not from records of how they were trained.

9 What training does a QMS auditor have to evaluate whether operator training is effective or not? It cannot be determined by a sampling audit or interview. Nor is it discoverable from traditional performance reviews.

10 How does an auditor determine if all the necessary documentation is available and suitable? The auditor would need to be an expert on each process to make this determination.

11 It would be common to check a sample of the available documentation to the organization's revision control logs. Again, such an audit would be inconclusive without checking a larger universe of controlled documents, not a sample. One revision error is too many.

As we were trained in auditing classes, traditional ISO 9001 auditing is a snapshot in time of the processes being audited. The scope of internal and external audits is determined ahead of time and the process owners are notified of the content and time of the audit. Unfortunately, this often gives the auditee the time to "clean up their act" by presenting an error free example or rehearsed process that may revert back to previously used shortcuts once the audit is completed.

We use non-scientific sampling methods to determine which process and procedures can be audited during the allotted time period. We accept the judgment of the auditor or audit team to determine compliance or nonconformance.

Internal audits typically plan to cover all elements of the QMS over a period of six months or a year. Process variability can sometimes be measured in mere hours or days. External initial certification or transition audits are still sampling audits and do not cover the entire QMS. External surveillance audits often plan to cover the entire QMS once in a three-year certification period.

While it is our cornerstone, traditional QMS auditing is grossly inadequate for determining day-to-day QMS compliance. Its efficacy in discovering foreseeable risk is almost nonexistent. Its ability to identify trends in conformance is limited to an almost accidental discovery of similar shortcomings of the QMS. Its ability to discover opportunities for improvement is determined by the competency and discretion of the audit team.

It is time to move beyond conformance auditing to forensic investigations focused on business process excellence and foreseeable risk. We must differentiate and then ingrain this into a company's culture so that it doesn't end up like the often started and abandoned standalone TQM or Six Sigma approaches.

Forensic Investigations

There are four types of forensic investigations:

a) Process Compliance – Determine compliance with documented procedures
b) Process Improvement – Assess opportunities for business process excellence
c) Risk Assessment – Assess opportunities to identify foreseeable risk
d) Expert Investigation – Conduct forensic incident investigations (Only for graduates of Course 5)

There are four subcategories of forensic investigators:

- Process Compliance Assessor – QM Level 2 Certification - Certified Process Management Expert (CPE)
- Process Improvement Assessor – QM Level 3 Certification - Business Process Excellence Master (BPXM)
- Foreseeable Risk Assessor - QM Level 4 Certification – Business Risk Avoidance Master (BRAM)
- Expert Investigator – QM Level 5 Certification – Business Management Systems Master (BMSM)

Process Compliance Investigation (PCI)

A process compliance investigation examines documented processes and their efficacy regardless of where they exist in the organization. Among the foundational tenets of the QM program is that undocumented business processes should not exist. Each process must have operational procedures and work instructions to the level of detail necessary to ensure process excellence and risk avoidance. They must also have appropriate roles and responsibilities for the process owners and operators.

Evolving processes and procedures to QM Level 1 standards is an iterative process of implementing an enterprise wide BMS. Each investigation may be used as a tool to form the evolution from QMS to BMS. These investigations are more than a snapshot interview of a designated department. They are in-depth investigations using proven quality tools such as FMEA, The New Seven Tools, RCA and creating Kaizen Events as needed for complex processes.

PC Investigations replace classic QMS audits. They include:

- Identification of the process to be investigated
  - Look at all interrelated processes that can be affected by associated risks.

- Collection of all applicable documentation related to the process
  - This should include a review of any Findings (and their follow-up) from previous audits.

- Identification of Roles and Responsibilities associated with the process
  - Look for any gaps in responsibilities and also for any overlapping responsibilities.
  - A RACI (Responsible, Accountable, Consulted, Informed) Matrix is an excellent tool to use here.
- A checklist of the planned outcomes of the process
- Concise criteria for evaluating the compliance of the process
  - An evidence-based evaluation of each operation within the process
  - An evidence-based evaluation of each role within the process
- Fact-based recommendations for immediate corrective action as warranted
- Fact-based recommendations for process improvements
- Appropriate metrics to indicate fulfillment
- Action plans to remove foreseeable risk

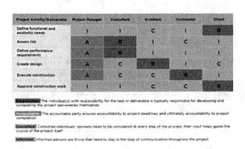

Project Activity/Deliverable	Project Manager	Consultant	Architect	Contractor	Client
Define functional and aesthetic needs	I	I	C	I	R
Assess risk	A	R	I	C	I
Define performance requirements	A	R	I	I	I
Create design	A	C	R	I	C
Execute construction	A	C	C	R	I
Approve construction work	I	I	C	C	R

Responsible The individual(s) with responsibility for the task or deliverable is typically responsible for developing and completing the project deliverables themselves.

Accountable The accountable party ensures accountability to project deadlines, and ultimately accountability to project completion.

Consulted Consulted individuals' opinions need to be considered at every step of the process, their input helps guide the course of the project itself

Informed Informed persons are those that need to stay in the loop of communication throughout the project.

Process Improvement Investigation (PII)

PII's are designed to assess the efficacy of a process with the intent to identify opportunities for measurable improvement and advance business process excellence. One approach to consider is a small, dedicated team (one or two Quality Professionals led by a Business Success Champion) that pulls in internal functional experts for a short period to conduct a deep dive into a specific business process. A Kaizen sprint can be used to find weaknesses and potential risks that need to be avoided, and then update the process based on the sprint findings. This reinforces the iterative processes stated in PCI above. Tips for successful investigations include:

- Start small – the greatest challenge to improving processes is feeling overwhelmed before you even start.
- Obtain stakeholder buy-in before you start and involve them during the investigation

These investigations include:

- Identification of the process to be investigated
- Collection of all applicable documentation related to the process
- Identification of Roles and Responsibilities associated with the process
- Current metrics of process effectiveness against business (functional level) Key Process Indicators
  - Time metrics measure how long a process takes to complete, such as cycle time, lead time, or throughput.
  - Cost metrics measure how much a process consumes or generates in terms of money, such as operating expenses, revenue, or profit.
  - Quality metrics measure how well a process meets the standards or specifications of its outputs, such as defect rate, error rate, or yield.
  - Customer satisfaction metrics measure how satisfied the customers or stakeholders are with the process or its outputs, such as satisfaction score, retention rate, or loyalty.
- Current competency evaluations of the process owners and operators
- An evidence-based evaluation of each operation within the process
- An evidence-based evaluation of each role within the process
- Specific process improvement recommendations with anticipated enhancements in metrics
- PII follow-up metrics monitoring at appropriate intervals

Risk Assessment Investigation (RAI)

The RAI should not be confused with risk management programs that may be implemented because of statutory or regulatory requirements or levied by insurance carriers. Organizations that have implemented risk management programs to ISO 31000:2018 may have invested considerable time and effort attempting to comply with ISO 31000 and may be mired in the complexities of that Standard. The point is NOT to add another compliance standard to an organization's infrastructure.

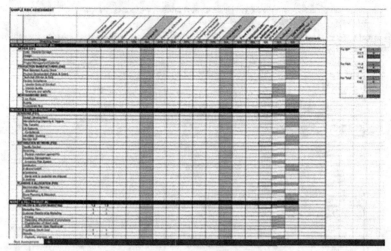

Figure 2 Sample Risk Management Matrix

Is completing a complex matrix such as depicted in Figure 2 really the solution to identifying foreseeable risk and implementing risk avoidance in your organization?

First, RA Investigations begin at the process level. Each process must be scrutinized in rigorous detail. We have many tools in our quiver to use diagnostically, such as DFEMA, PFEMA, House of Quality, Ishikawa Diagrams[5], etc. There is no need to construct complex matrices of risk management to diagnose foreseeable risk in a process or begin new risk avoidance initiatives. When we have completed a RAI on each process, the operational procedures, work instructions, metrics and roles must be revised to ensure the investigation results are implemented. As we link individual processes that are immune from foreseeable risk, we grow a tree that eventually addresses all areas of organizational risk. RAI's must start at the lowest level in the production process. Following are elements of a good RAI:

- Identification of the process to be investigated
  - Identification of interrelated processes that can be affected by associated risks.
  - Gathering of historical audits and Findings as a baseline for the current audit. Are there any repeat Findings? If so, were associated risks properly defined?
- Collection of all applicable documentation related to the process.
- Identification of Roles and Responsibilities associated with the process.
- Identification of any risks previously discovered in the process.
  - These can come from internal audits, customer feedback and from input from your operators and technicians who do the job of producing your goods and services.
- Identification of all potential risks that the process can generate either directly or indirectly.
- Current competency evaluations of the process owners and operators.
- An evidence-based evaluation based on "what could possibly go wrong?"
- An evaluation of how customers can potentially misuse the product or service.
- Identification of potential foreseeable risk within the definitions of duty of care and standard of care.
  - Duty of care is the legal obligation of a person or organization to avoid acts or omissions that could likely cause harm to others. Standard of care is the level of reasonableness expected in the circumstances.

---

[5] See Figure 4

- Specific process improvements that will eliminate the root cause of the foreseeable risk.
- Follow-up metrics monitoring at appropriate intervals.

RAI's are part of a perpetual enterprise-wide program of identifying foreseeable risk. They can also be used to diagnose risks within statutory and regulatory processes.

ISO 9001:2015 REQUIREMENTS	FORESEEABLE RISK
3 Terms and definitions	Define Foreseeable Risk
4 Context of the organization	Do No Harm
4.1 Understanding the organization and its context	Incorporating the Risk Avoidance Model into the BMS
4.2 Understanding the needs and expectations of interested parties	
4.3 Determining the scope of the quality management system	
4.4 Quality management system and its processes	
5 Leadership	Modeling Risk Avoidance
5.1 Leadership and commitment	No Defect Ever Reaches a Customer
5.2 Policy	
5.3 Organizational roles responsibilities & authorities	
6 Planning	Implementing Risk Avoidance
6.1 Actions to address risks and opportunities	Incorporating the Tenets of Foreseeable Risk into the BMS
6.2 Quality objectives and planning to achieve them	
6.3 Planning of changes	
7 Support	The Foreseeable Risk Infrastructure
7.1 Resources	Incorporating the tools of Foreseeable Risk into the BMS
7.2 Competence	
7.3 Awareness	
7.4 Communication	
7.5 Documented information	
8 Operation	Implementing Defect Free Processes
8.1 Operational planning and control	Incorporating Risk Avoidance in all Process Steps
8.2 Requirements for products and services	
8.3 Design and development of products and services	
8.4 Control of externally provided processes products and services	
8.5 Production and service provision	
8.6 Release of products and services	
8.7 Control of nonconforming outputs	
9 Performance evaluation	Risk Assessment
9.1 Monitoring measurement analysis & evaluation	Evaluating Risk Exposure
9.2 Internal audit	
9.3 Management review	
10 Improvement	Foreseeable Risk
10.1 General	Ongoing Risk Avoidance
10.2 Nonconformity and corrective action	
10.3 Continual improvement	

Figure 3 Modeling Risk Avoidance in ISO 9001:2015

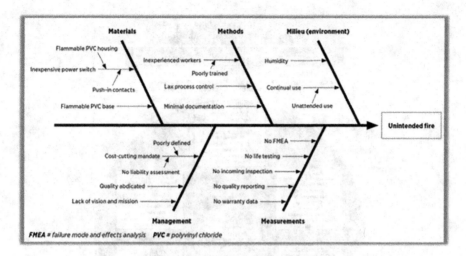

Figure 4 Ishikawa Method for Forensic Investigations

Figure 4 Process Excellence and Risk Evaluation Tool

Expert Investigation (EI)

An EI is launched in response to an identified critical nonconformity or to investigate behaviors or omissions leading to organizational liability. These are conducted in-house by a senior level investigator who has skills that can include those of an expert witness in products liability and organizational negligence.

The incident typically dictates the depth of conduct of the investigation. In the case of a critical defect being identified within a process, or series of processes, the EI is conducted much the same as a RAI. If the EI is the result of a legal action against the organization, the Expert becomes part of a legal defense team. In that case, the attorney of record sets the agenda for the defense. The Expert is tasked with ascertaining that the organization exhibited an acceptable duty of care in delivering the product or service to the stream of commerce.

In the scenario where the Expert is testifying for the plaintiff, the attorney representing the injured party sets the agenda for the plaintiff's case. The EI then focuses on proving inappropriate standard of care by the defendant. As part of the plaintiff's team, the Expert must continually compare their investigations to the other experts in the case.

A typical litigation EI includes:

- Read all case filings and counter filings
- Studying production of complaints and cross filings with the court
- Studying evidentiary filings produced by both parties
- Determine the duty of care that can be correctly imposed on the defendant (such as ISO9001, AS 9100 or IATF 16949)
- Determine if the defendant is certified to a credentialed management standard
- Request discovery of the defendant's quality manual and any subordinate documentation they used to claim compliance to a quality standard
- Follow the steps of a PCI to determine compliance with the applicable standard
- Create a report that is focused on how the defendant failed to comply with the standard or with their own process documentation
- Ensure the report is congruous with the testimony of the other experts

This methodology does not assume the credentials necessary to be an expert witness that offers themselves to attorneys. This must be combined with experience. Obtaining the first few assignments to create a CV that includes casework is much like breaking into any new career. The skills to present yourself as competent to the attorneys requires creating the case that your quality management experience is the most valuable part of your service. The other compelling sales tool is that you will be assessing duty of care and standard of care from an authoritative perspective, not from an opinion, while other experts deal with causes and origins. It is often difficult to get an attorney to look beyond their paradigms to understand the difference and value of a FI.

- Establishing Investigation Objectives
- Internal, second party and third-party investigations share the objectives of both conformance to a standard and the objectives of each of the four types of investigations described above. The organization needs to create a comprehensive requirements document for each investigation.

The requirements document must include:

- The specific scope of the investigation
- The needed outputs
- The skills of the investigators
- The competence expectations of those being interviewed
- Identification of observed and foreseeable risks
- Identification of opportunities for process excellence
- Action items from previous investigations
- New action items from this investigation
- Evaluation of the suitability of the investigation

Determining Risks

There are innate risks in conducting FI's. Different techniques can be used to determine and prioritize risks. There are traditionally two types of risk assessment. The first is Qualitative – focusing on the perceptions about the probability of a risk occurring and its impact. These are usually represented in scales such as "low-medium-high". The second technique is Quantitative – focusing on measurable data to calculate the probability and impact of identified risks.

Regardless of which technique you use, the basic elements you need to look at include:
- Impact on those being interviewed
- Impact on resources
- Impact on business operations
- Competence of the investigators
- Effectiveness of oral and written communications
- Document integrity.
- Establishing the FI Program
- Implementing the FI Program
- Monitoring the FI Program
- Continual Improvement

Each of these must be addressed in creating the requirements document. Most companies use a qualitative risk technique as it is faster and easier to conduct than a quantitative analysis; however, it is subjective in nature and doesn't provide the detailed data that would be necessary for legal actions.

Other FI Tools

All companies certified to a Standard; be it ISO/AS 9001, OSHA, National/ International FAA, Nuclear Standards, or any number of others, conduct internal audits in order to maintain compliance. This is required to maintain certification but most stop at simple compliance. This Guidance Document is not meant to replace any of the current Standards and their auditing practices but instead, offers techniques to enhance them and in turn, allow your company to improve existing processes by looking at ways to increase efficiencies and reduce or remove nonconformances. By utilizing the audit tools outlined here, you can often find hidden improvements. Here we present four auditing tools that can help a company become more efficient, profitable, and competitive. These are cyclical and can be used independently in a Kaizen Event or as a group as part of an overall plan to improve your business.

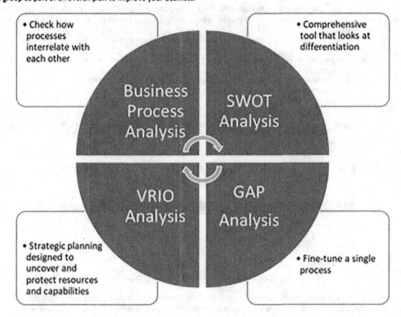

Business Process Analysis (BPA)

With a standard internal audit plan, companies typically prepare an annual plan that looks at individual processes and procedures but often don't look at how these interrelate with upstream and downstream processes. A BPA is a method designed to help you review the workflow throughout your entire enterprise.

There are 5 steps for a BPA:
- Review the big picture. Ask how your processes support business goals. Do you have a Vision Statement? What are your long-term goals? Do they support your company's mission?
- Interview the people who do the work. Process owners are important stakeholders in your company's success but don't forget the floor/cell level technicians. The touch labor employee often knows more about the process than a supervisor or manager and usually has good recommendations for improvement.
- Analyze individual process steps and associated KPIs. Is there an overlap between an up/downstream process that can be eliminated or combined to help prevent confusion about who is responsible for a sub-step?

- Identify opportunities for improvement. Look at your technology. Is there a process or process step that can be automated? Is there a way to improve communication between process and sub-steps? Should a process or step be moved to make an interrelated process more efficient? Have you identified all resources needed? See above and ask your workforce for recommendations.
- Make changes as appropriate. This is where you sum up everything you've learned from audits and GAP, SWOT, and VIRO analyses.

SWOT Analysis

This is a familiar tool used to determine Strengths, Weaknesses, Opportunities and Threats within your business by looking at internal and external factors that can affect your company's agility and competitiveness. As the acronym suggests, there are four areas to review.

- o Strengths – What advantages or characteristics of your business give you an advantage over your competitors?
- o Weaknesses – What characteristics of your business or project place you at a disadvantage relative to your competitors?
- o Opportunities – What elements of your business can you exploit to its advantage?
- o What elements in your business or project could cause trouble?

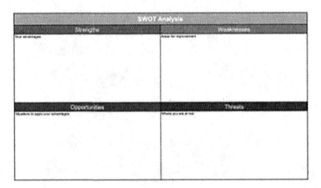

GAP Analysis: This is used to help identify your present state (where you are now) with your target or end state (where you want to be). It helps determine whether you are meeting expectations and using resources such as time, money, employees, technology, and infrastructure effectively. It can be used at a macro (strategic) or micro (process/step) level to categorize gaps that you need to close. After conducting a GAP analysis, you should have a better idea of where to focus your resources. There are typically four steps to a GAP analysis:

- Identify your current state. Ask yourself, "Where are we at now?" Make a list of what is and what is not in the area you are studying. Collect qualitative information such as processes and methodologies used to produce your products and services and quantitative information that can be counted and measured.
- Identify your proposed end state. This is your desired state, future target or stretch goal. To get to your end state, as you are looking at your end state, also review your Vision, Mission and Goals to see if any adjustments are needed. Make sure that your desired end state is reasonable and has a realistic timeframe to get there.
- Identify the gaps. Once you know where you are and where you want to be, try to determine why there is a gap. Take an honest look at how the gap occurred by looking at the specifics of your business, its processes and your use of resources.
- Bridge the gap. Ask what needs to happen to get to your goals. Develop a plan with detailed, unambiguous action items. Make sure these are realistic by reviewing how you will use your resources. Finally, ensure that you follow through with your plan and celebrate your successes along the way.

VRIO Analysis

This is a relatively new tool for the Quality Master that will help your senior level executives make business decisions for developing and maintaining a sustained competitive advantage. This directly ties to Clause 08 OPERATION of the BMS 9001:2023 Guidance Document. It is a strategic planning method used to evaluate internal resources within your company. VRIO stands for:

- The Value of your resources. Do you offer a resource that is valuable to your customers? Can you exploit an opportunity or neutralize an external threat with this resource?
- The Rarity of your Resources. Is there a resource that you need that is difficult to obtain or in short supply? Do you have enough inventory on-hand or a plan to obtain it? A GAP analysis is a good tool to use here.
- The Imitability of your product and its resources. Do you control scarce resources or capabilities? Can your product or service be easily imitated or duplicated? If it is rare or easily obtainable, it's likely your competitors will try to duplicate it or create a substitute.
- Does your organization have the internal structures, systems, and processes to exploit your product and services advantages? Does your management make informed decisions that align with your company's strategies?

Simply possessing the resources isn't enough to have sustained competitive advantage. Once you have looked at the value, rarity, and cost to competitors to duplicate your products and services, you then need to organize the company to exploit your resources.

Summary Of The VRIO Model

Is it valuable?	Is it rare?	Is it hard to imitate?	Is the firm organized around it	What is the result
✗				Competitive Disadvantage
✓	✗			Competitive Equality
✓	✓	✗		Short - Term Competitive Advantage
✓	✓	✓	✗	Unused Competitive Advantage
✓	✓	✓	✓	Long - Term Competitive Advantage

The Balanced Scorecard (BSC)

A Balanced Scorecard (BSC) is a strategic management performance metric used to identify and improve various business functions and their outcomes. It measures past performance data and provides organizations with feedback on making future business decisions.

A BSC can be used as a standalone tool for management reviews or to summarize the results of a Forensic Investigation. After you have conducted your process compliance, improvement and foreseeable risk investigations, and have identified or updated your business and functional KPIs, a scorecard can be used to track your progress towards achieving your Vision, Mission, and Goals.

Process/Owner	Scorecard Category	Higher level Objective Supported	Metric	Target	Goal	Company Average To Date	Action Plan	Frequency Check
Budget Execution Management	Financial	Demonstrate accountability and efficiency	Actual vs planned expenditures	Below Budget 4%	Below Budget 4%		Set Target/Goal	Quarterly
			Revenue	TBD	TBD		Set Target/Goal	Quarterly
			Operating Expense Ratio Total Operating Expense / Revenue	4% < half years OER	4% < half years OER		Set Target/Goal	Annual/Quarterly
			Cost vs Revenue	TBD	TBD		Set Target/Goal	Annual/Quarterly
Contract Performance Execution and Service Managers	Customer	Superior performance in the eyes of the customer to increase opportunities for selection on future contracts	Customer Score - Quality	Satisfactory	Exceptional		This is dependent on the type of product or service produced. If you request input from your customer, these scores are valid. If no customer input is received, conduct an honest, critical internal evaluation based on developed metrics. Develop stretch goals as your metrics improve.	Annual
			Customer Score - Schedule	Satisfactory	Exceptional			Annual
			Customer Score - Cost	Satisfactory	Exceptional			Annual
			Customer Score - Management	Satisfactory	Exceptional			Annual
			Customer Score - Other	Satisfactory	Exceptional			Monthly
			CAR/all PARs issued by Customer	# CAR/all PARs 0 - 2 PAR/all PARs 0				Monthly
			Call-out Completion Time to ECD	>% completed on time or ahead of ECD	>% completed on time or ahead of ECD		Set Target/Goal	
Employee Performance & Staffing Managers, HR	Learning and growth	Complete employee performance management objectives and development plans on time and use to improve performance	Objectives submitted on time	> 90%	90%		Continue as usual	Quarterly
			Mid year reviews completed on time	> 90%	90%		Continue as usual	Quarterly
			Significant item(s) on all development plans executed	> 90%	60%		Survey Managers - Develop and send out Training Needs Assessment and then develop training where there are gaps in development/skills	Quarterly
			Assigned training completed on time	> 85%	> 90%		How do you track training? Is it effective, are assigned classes/training completed on time?	Quarterly
			Employees understand what training is upcoming and what is overdue	> 90%	90% Continue		Track 30 day upcoming & overdue training for their area of responsibility	Monthly
			Average years of service	Increase year over year	Increase year over year		Set Target/Goal	
			Employee Turnover	< 5%	< 5%		Set Target/Goal	Monthly
Quality Performance Senior Functional Mgrs, Quality Mgr	Internal Business Processes	All metrics at or above Target with an improvement trend over time	Audits developed, scheduled and completed on time	90%	90%		Benchmark other companies	Monthly
			ISO 9001:2015 compliance	100%	100%			Annual

A BSC looks at much more than process efficiency or product defect rates. It allows you to show structure within your overall company strategy. While each internal department may track individual KPIs differently, a BSC allows management to organize these in an easy-to-understand format. Management can see where they are now, where you want to be in the future, and how you plan to get there. It is flexible enough for even the smallest company to create and can be used by companies that have multiple sites for comparison. The key to a successful BSC is in the last column – the frequency check. Realistically reviewing your status on a periodic basis forces you to make adjustments on a more regular basis which in turn helps you stay healthy in an ever-changing business environment.

WARNING

Do not include any of these FI tools in your quality manual, web site or other public-facing documents. They could be used against you in any lawsuits you might face.

Appendix E: Forensic Investigation Checklist for ISO 9001:2015

Forensic Investigations (FI) are more robust and beneficial than traditional management systems and process audits and will assist your transition from a QMS to an enterprise-wide Business Management System (BMS). The following is a checklist correlated to the ISO 9001 clauses that will help you ensure not only compliance but also help you identify foreseeable risks, how to avoid risks instead of mitigating them, and how to ensure opportunities for process excellence are identified and put in place. Note that items in blue font refer to compliance, opportunities for improvement, risk, and opportunities for process excellence as outlined in the Forensic Investigation Handbook and BMS 9001:2023.

Clause 4 Context of the organization: This first clause introduces two sub-clauses relating to the context of the organization: (1) understanding the organization and its context and (2) understanding the needs and expectations of interested parties. Together, they require that you determine the issues and requirements that can impact the planning of the Quality Management System (QMS). In addition, the scope of the QMS/BMS and the QMS/BMS processes along with their applicability and interactions, need to be determined.

- Identify your Mission, Vision, Goals, and Scope of your BMS and the Key Process Identifiers (KPIs) and Strategy for attaining them.
- Determine and define internal and external stakeholders.
- Identify and document all statutory, regulatory, and customer requirements.
- Work with management and functional/department owners to identify external and internal issues that affect your Vision, Goals, and organizational KPIs, not quality metrics.
- Determine how you monitor and review your KPIs and how they affect your stakeholders.
- Verify that all process inputs/outputs, methods for monitoring and measuring processes, and resources needed to achieve your KPI goals are in place.
- As you conduct your FI, look for associated risks and opportunities (see Clause 6 further on) for improvement to your BMS.

Clause 5 Leadership and commitment: This clause requires that your top management demonstrates leadership and commitment with respect to the BMS. In addition, top management is required to demonstrate leadership and commitment with respect to customer focus.

- Interview top management and ask how they take accountability for the effectiveness of your BMS.
- Review your Quality Policy and objectives to ensure they are current and relative to your company's mission.
- Interview employees to see if they understand the importance of effective quality and ask how this is communicated.
- Interview management and employees to determine if they understand who your customers are, their needs, and if they understand any statutory, regulatory, and safety requirements.
- Check to see if organizational roles, responsibilities, and authorities have been established and are understood by all.

Clause 6 Planning: This clause talks about the planning for the QMS/BMS, where your company needs to consider the issues and requirements referred to in Clause 4 and determine the risks and opportunities that need to be addressed. In addition, this section covers the quality objectives that will need to be established for the relevant functions and the plans to achieve them determined.

- Review your BMS to ensure that it can achieve its intended results. Look for desired effects within each process, how the processes interrelate, how undesired effects (risks) are identified and removed, and how improvements can be put in place.
- Review your KPIs with management and functional/department owners. Ask if they are consistent with the quality policy and how objectives are measured.
- Ask how KPIs are measured, how often they are reviewed, and what happens if they do not meet their goals.
- If changes to your BMS are needed, ask how they are identified and how they are carried out.
  - Are all resources for the change identified and available?
  - Are risks identified and planned for?
  - How are opportunities for improvement considered and planned for prior to putting them in place?
- Interview process owners to determine if their quality objectives are relevant and are in place with customer satisfaction in mind.

Clause 7 Support: This clause requires that your company determine and provide the resources needed to establish, implement, maintain, and continually improve the QMS/BMS. It includes personnel, infrastructure, environment for process operations, monitoring and measuring, and organizational knowledge.

- Interview management and department/functional owners to determine if appropriate resources are available and if a plan is in place to implement, maintain, and continually improve the BMS.
- Verify that your company's employees are trained and can meet statutory, regulatory, and safety requirements for their processes.

- Look to see that the infrastructure (buildings, utilities, equipment, software/hardware, IT, and transportation) is sufficient to produce your products and services.
- Ask employees if their work environment is sufficient for the operation of their processes to achieve product requirements.
- When monitoring and measuring the conformity of products and services, ensure that equipment (e.g., Test, Measurement, and Diagnostic Equipment) is available and properly maintained. Review any documentation for evidence of training on the equipment to verify calibration and its condition during use.
- Ask employees what the process is for equipment found out of calibration or damaged. Determine if there is a process to backtrace the use of the equipment and if a product recall process is in place if needed.
- Confirm that knowledge necessary for the operation of processes and to achieve conformity of products and services is in place.
  - Is formal training available and documented?
  - for organizational knowledge, is a plan for information such as intellectual property and lessons learned?
- Ask how the competence of personnel (e.g., formal education, training, and experience) to do their job is documented. Ask what actions are taken to acquire the necessary competence and how it is evaluated.
- Interview employees for their awareness of the quality policy, objectives, how they contribute to quality and company excellence, and the implications of not conforming to processes.
- Ask how internal and external requirements and concerns are communicated.
- Documented information—look for the following:
  - How are processes, procedures, and work instructions maintained?
  - Is product safety taken into consideration, and how is this documented?
  - How is documented information created and updated?
  - How is it controlled—distribution, stored, accessed, retained, and archived?
  - How is external documentation (e.g., customer specifications) controlled?
  - Is documented information retained as evidence of conformity and protected from unintended alteration? How is this done?

Clause 8 Operation: This clause requires that your company plan, implement, and control the processes required for the BMS to implement the actions to address risks associated with operational processes. Operational planning and control include systems for configuration management, product safety, prevention of counterfeit and unapproved parts, and installation of approved parts. In addition, systems for customer-related processes, design and development, control of external providers, control of production and service provision, including identification and traceability, preservation of products, and control of nonconforming outputs, are required.

- As you conduct your Forensic Investigation, look to see how your company plans, implements, and controls the processes necessary to meet customer, statutory/regulatory, and safety requirements for your products and services:

- After finishing production, how do you accept products and services? Are adequate controls in place? Who approves them?
- Are the resources needed to achieve conformity available, controlled, and properly documented and retained?
- How do you identify and control planned changes and review for unintended consequences? What actions are taken when a risk is identified?
- Does your company ensure that outsourced processes are properly controlled?
- Have you planned, implemented, and controlled the processes needed for product and service safety?

- Ask management how they communicate with customers for:
  - Information relating to products and services.
  - Contract inquiries.
  - Customer feedback, including complaints.
  - Handling and storage of customer property.
  - Specific requirements for contingency actions if needed.
- How are product and service requirements defined, and can these defined requirements meet claims for products and services offered?
- To ensure that you can meet requirements, conduct a review of your company's commitment to/for:
  - Review of specified customer requirements for delivery and post-delivery activities.
  - Review of any customer requirements not specifically stated but are essential for the intended use.
  - Any statutory, regulatory, and safety requirements.
  - Contract or order requirements differing from those previously expressed.
  - How any changes to customer requirements are handled and documented.
- If your company designs and/or develops the product, does your company consider:
  - The stages and controls during the design and development effort? To Understand the difference between verification and validation and the controls needed for each? Are these defined and followed?
  - To the level of control expected by customers or other interested parties.
  - Necessary documented information to confirm design and development requirements are met.
  - Applicable statutory/regulatory requirements?
  - Are there standards and codes of practice that your company is committed to implementing at all levels?
  - What are the potential consequences of failure due to the nature of your products and services?
  - Do you identify, review, and control any changes to ensure no adverse impacts on conformity to requirements?
  - Is documented information retained for:
    - Design and development changes.
    - Results of reviews.

  - – Authorization for changes.
  - – Actions taken to prevent adverse impacts.
- Ask if/how your company ensures that externally provided (vs. procured) processes, products, and services are controlled.
  - Are external (supplier) products or services provided directly to your customers under your name or contract? How do you control the conformity of these products or services?
  - How do you evaluate suppliers who provide products or services directly to your customers?
  - Do you conduct periodic reviews of suppliers who provide direct products or services to your customers? Do you consider potential adverse impacts, and if so, what steps do you take when a product or service does not meet requirements?
  - Do you maintain appropriate documented information for the results of evaluations and monitoring of the supplier's performance?
- Information for external providers:
  - Do you maintain an Approved Supplier/Vendor List (AVL)? How are suppliers approved, and who has the authority to approve (e.g., procurement only or procurement + quality, or some other mix, such as engineering input)?
  - How do you flow down specifications and quality requirements to the supplier? How is this documented?
  - How do you evaluate suppliers? How often?
- Product and service provision:
  - Do you have controlled conditions for the availability of documented information defining characteristics of the products and services and activities to be performed to ensure compliance?
  - How do you control the availability and use of monitoring and measuring devices? How do you ensure calibration is up to date and that no uncalibrated tools are used?
    - – How do you quarantine uncalibrated or broken devices?
    - – Do you have a process to trace tools found out of calibration to products or processes they may have been used on before out of calibration was discovered? How is this documented?
    - – Do you have a product recall process if needed?
    - – Are personnel trained and competent to use monitoring and measurement tools? Is this documented?
    - – Do you have a process or action to prevent human errors?
- Property belonging to customers or external providers:
  - How do you identify, verify, and protect external property including software and firmware?
  - When external property is damaged or found to be uncalibrated, how do you document this and report it to the customer or external provider? How is this property quarantined?
- Preservation:
  - Do you include identification, handling, packaging, storage, transmission or transportation, and protection as part of preservation?

- Post delivery:
  - Does your company meet requirements for post-delivery activities associated with the products and services?
  - When determining your post-delivery activities, do you consider the:
    - Statutory, regulatory, and safety requirements?
    - Potential undesired impacts associated with your products or services?
    - Nature, use, and intended lifetime of the products or services?
    - Customer requirements and feedback?
- Control of changes:
  - How does your company review and control changes for production or services to ensure continuing conformity with specified requirements?
  - Do you identify the persons authorized to approve production or service provision changes?
  - Do you retain documented information describing the results of the review of those changes, the personnel authorizing the change, and any necessary actions?
    - Is there a process in place to identify adverse impacts?
    - What actions are taken when adverse impacts are identified?
- Release of products and services:
  - Do you validate that your products and services meet your and your customer's requirements prior to release?
  - How is this documented? Are acceptance criteria included, and do you ship certificates of conformance (or similar) with your product?
- Control of nonconforming outputs:

  - How do you ensure that nonconforming outputs during production are identified and controlled?
  - How do you document nonconformities and any corrective actions taken prior to release to the customer? How long is this documented information retained?
  - How do you take corrective action for products or services found nonconforming after release to the customer?
  - After making corrections, how do you verify the requirements are met?

Clause 9 Performance evaluation: This clause requires that your company plan, implement, and control the monitoring, measurement, analysis, and evaluation processes. Performance evaluation includes systems for the evaluation of customer satisfaction, analysis and evaluation of data, internal audits, and management review aimed at improved quality performance and an effective QMS/BMS.

- How do you determine what needs to be monitored and measured?
  - When is monitoring and measurement performed?
  - Are results documented? How are they maintained/controlled, and for how long is this retained?

- Do you evaluate the quality performance of your QMS/BMS?
  - How is this done?
  - How is this documented?
  - How often is this reviewed, and by whom?
  - Do you request customer input? If so, how?
    - Does this include such things as customer surveys, complaints, warranty claims, etc.?
  - Do you track on-time delivery?
- Is an internal audit process in place?
  - Is an internal audit plan developed and followed?
  - How do you determine effectiveness?
  - How do you report findings?
    - How are these reported?
    - How do you close them?
    - Do you conduct follow-ups to confirm closure?
  - Do you identify risks and opportunities for improvement?
  - Do you retain appropriate documentation?
- How do you track customer satisfaction, needs, and expectations?
- Do you conduct management reviews of your QMS/BMS at planned intervals?
  - Do top management and functional owners attend?
  - Do you discuss inputs and outputs such as KPIs, on-time delivery, and customer satisfaction?
  - Do you review the effectiveness of actions taken to address risks and opportunities for improvement?
  - Is safety (product, employee, environmental, etc.) on the agenda?

Clause 10 Improvement: This clause requires that your company determine and select opportunities for improvement and implement the actions needed to meet customer requirements and to enhance customer satisfaction. The improvement process includes systems for nonconformity and corrective action and for continual improvement.

- How do you determine and select opportunities for continual improvement?
- What actions are taken to meet customer requirements and improve customer satisfaction?
- How are nonconformities and corrective actions addressed?
- How are changes to your QMS/BMS made?
  - Who has authority?
  - How is effectiveness determined?

Appendix F: BMS 9001:2024

BMS 9001:2024

(Free abridged version)

A BUSINESS PROCESS MANAGEMENT
GUIDANCE DOCUMENT

This document is in no way connected to the International QMS Standard ISO 9001:2015, the International
Organization for Standardization, nor any other organization. It follows
the numbering convention of ISO 9001:2015 and includes the intent of that Standard in its respective clauses.

Author: Tom Taormina, CMC, CMQ/OE, FBP, QM
Senior Contributor: Steven Garner, QM

Revision 1 | March 2024

–CONTENTS–

–FOREWORD–

Since the release of ISO 9000 in 1987, the Standard has primarily been used as a compliance and conformance document. For many organizations, it is implemented to comply with customer requirements for an approved supplier. Others have implemented it to meet organizational directives or mandates. Some have offered it to senior management as a tool to minimize defects and customer dissatisfaction. By design, it is, in fact, a conformance standard.

Precious few have implemented it as an instrument of business process excellence. This evolutionary methodology will retool quality management systems into an enterprise wide business management system. It will supplement conformance and compliance mandates with process excellence tools; to create return on investment (ROI), rather than overhead expenses.

It will also replace traditional risk management with risk avoidance. That is, instead of a scheme to manage the hazards of risk, it employs process management tools to avoid risk at every stage from initial design through delivery to the customer.

The International Organization for Standardization in Geneva is starting to revise ISO 9000, it will most likely take several years for the ISO family of Standards to update, implementing BMS 9001:2024 will allow you to get a big head start with a new benchmark for the future. By following the tenants of this document certified companies will by design meet all requirements of the Standard when they are audited by third-party assessors.

–INTRODUCTION–

1 GENERAL

Adopting a business management system is a strategic decision for an organization to achieve enterprise-wide process excellence and risk avoidance. The potential benefits to an organization implementing a business process management system (BMS) based on ISO 9001 are:

- The ability to consistently provide products and services that meet customer and applicable statutory and regulatory requirements
- Creating a goal of 100 percent customer satisfaction
- Creating a goal of 100 percent risk avoidance
- The ability to demonstrate conformity to specified management systems requirements.

As with the ISO Standards, the following terms are used:

- "Shall" indicates a requirement that will be met
- "Should" indicates a recommendation
- "May" indicates a permission is needed
- "Can" indicates a possibility or a capability

2 BUSINESS MANAGEMENT SYSTEM PRINCIPLES

These principles have been developed in a collaboration among eminent quality professionals to enhance the implementation of ISO 9001:2015 and to focus on business process excellence and risk avoidance.

These BMS principles include:

- Vision - Create products and services of exceptional value. Your vision needs to be specific enough for your people and other interested parties to understand where you want your company to evolve and how it stands out from your competitors.
- Mission—Create a sustainable and scalable organization that executes the vision created by your business leaders.
- Values—Success is the inevitable result of achieving the vision based on your defined business ethos.
- Leadership—Enlightened visionaries who are great communicators and inspire their employees to take ownership of their work through empowerment and engagement.
- Process—All activities are a series of interrelated processes. All stakeholders must understand where and how their roles interact with upstream and downstream processes.
- Boundaries—Everyone operates with an agreed-upon standard of accountability.
- Metrics—All business process decisions are based on comprehensive, measurable data, derived from Key Process Indicators (KPIs) put in place at appropriate points inside your process flow.
- Consistency—Quality, reliability, value, timeliness, and profit are all achieved concurrently.
- Achievement—Everyone wins. No exceptions. Society is enriched by the people and accomplishments.

Effective business process management sunsets *risk management*, replacing it with *risk avoidance* as a fundamental tenet of the organization.

3 THE PROCESS APPROACH

The process approach involves the systematic definition and management of processes, and their interactions, according to your quality policy and the strategic direction of your organization.

Management of the processes and the system can be achieved using the 6A Model.

The Six A's of Process Optimization

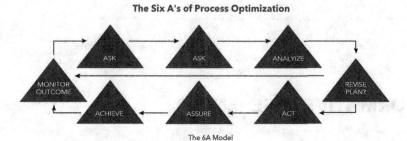

The 6A Model

The 6A Model is an enhancement of the Plan, Do, Check, Act (PDCA) model referenced in ISO 9001:2015, and in most training for implementing the Standard. It includes critical missing steps, such as "ASK" what the needs are before any planning. The 6A Model is a challenge to your concepts of process analysis. See Appendix C.

–REQUIREMENTS–

1 SCOPE

This document contains the essence of the requirements of ISO 9001:2015 enhanced to elevate that Quality Management System (QMS) to the BMS 9001:2023 Business Management System Guidance Document. Organizations certified to ISO 9001:2015 can continue to maintain their certifications even if they replace their QMS with BMS. External auditors will continue to audit to ISO 9001:2015.

All the requirements of this Standard are universal and applicable to any organization, regardless of type, size, or the products and services it provides.

2 NORMATIVE REFERENCES

ISO 9000:2015 Quality Management System -- Fundamentals and vocabulary

3 TERMS AND DEFINITIONS

Terms and definitions are included in Appendix A.

4 CONTEXT OF THE ORGANIZATION

4.1 Understanding the organization and its context

Clause 4.1 compels the organization to look objectively at internal and external issues that can affect its strategic objectives. By systematically identifying every issue, a matrix or process flowchart can be developed to ensure that every critical factor is identified and continually evaluated, controlled, and improved.

In understanding the organization, review traditional overhead risk management tools and requirements. It is imperative to discard non value-added processes.

Most organizations are target-rich in foreseeable risk at all levels. In developing a culture of risk avoidance, you will build the tools and tenets of foreseeable risk into every internal and external process.

Customer interactions are discussed in other clauses, but it is critical to include customers in examining organizational external issues. Foreseeable risks with customers must include assessing how customers can use your products and services badly or for the wrong purposes.

Ask: Have you considered the relevant issues—internal and external—that can affect your ability to achieve the goals of your business system? How is your customer involved, and what are your responsibilities after you have delivered your product or service?

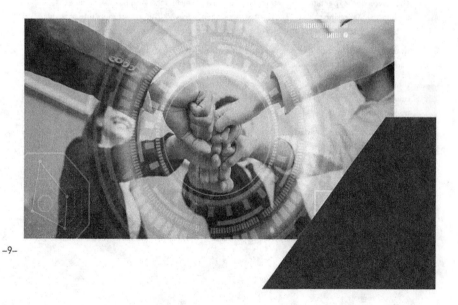

4.2 Understanding the needs and expectations of interested parties

Clause 4.2 requires that the organization evaluate the needs and expectations of all "interested parties." It ensures that everyone, from your people to customers, suppliers, and shareholders, are in the loop, and that a channel exists for effective communications. With influencers and interested parties determined and codified, the organization can more effectively determine the true scope of your BMS. All these key inputs are necessary to build an effective, process-based BMS.

The needs and expectations of interested parties are key in achieving business process excellence. Assessing the needs of customers is mandatory. Turning customers into referrals, instead of liabilities, is a critical step of understanding their needs.

Understanding the motivations of your people will turn them into accountable process operators and owners, and it will immediately enhance key business metrics.

Understanding the needs of shareholders will help transition from quality management to whole business process management. You will better understand why governmental agencies and nongovernment agencies impose specific requirements and standards for your industry.

In the context of risk avoidance, all interested parties must scrutinize foreseeable risk. Traditional risk management programs may prescribe risk-analysis methods, but you need to replace them with a methodology that eliminates risk in the first place, rather than simply managing it.

Ask: Does your company monitor and review the information about these interested parties and their requirements?

4.3 Determining the scope of the quality management system

Clause 4.3 is designed to ensure that senior management has formally defined processes with outcomes that can be measured and continually improved. The process owners, suppliers, and customers also have a baseline from which to interact with the organization.

Traditionally, determining the scope means codifying a framework based on 4.1 and 4.2. This framework must include all the requirements of the Standard. It includes processes that are mandatory for an effective BMS. It includes a scope statement that is required for certification.

Similarly, the defined processes must identify the areas of foreseeable risk in each process and procedure.

Turning a QMS requirement into a BMS framework includes all key business process requirements. This is a change in basic assumptions, from process management and metrics to a framework of cohesive and value-add content within the scope.

4.4 Business management system and its processes

Clause 4.4 is the imperative to evolve a traditional QMS into a BMS. In recasting the quality policy and objectives into vision, mission, and values, you must identify the immutable imperatives of your business processes.

You will need to determine:

- Inputs and outputs expected from your business processes
- Criteria and methods, including measurements and related KPIs needed to ensure the effective operation and control of the processes
- Resources needed for successful process execution

To include the concepts of "quality as a profit center1," each process must be evaluated for return on investment and effective interaction with other processes. This definition is the evolution from a QMS to BMS. But that term must be customized to your environment. For instance, "profit center" has other measurable metrics in a not-for-profit organization. In defining the processes and their interactions, include the tenets of risk avoidance in your processes.

---

1 See Appendix D.

5 LEADERSHIP

5.1.1   Leadership and commitment

ISO 9001:2015 Clause 5.1.1 lists 10 requirements that are intended to ensure that operational leadership is fully invested in the BMS, and that they embrace it as a business process imperative. This clause defines leadership and commitment as enlightened senior executives whose goal it is to drive the BMS principles enumerated in section 2 on Business Management Principles.

Sub clause 5.1.1 d) requires top management to promote risk-based thinking. A more effective requirement is to establish and model a culture of risk avoidance. Create guidelines, training, and processes for identifying foreseeable risk avoidance and removing it. This is covered in detail in Appendix B.

Ask: What does risk-based-thinking really mean?

5.1.2   Customer focus

This requirement in ISO 9001:2015 covers three areas of maintaining statutory requirements, assessing risk avoidance, and enhancing customer satisfaction. Though these requirements are valid and necessary, the commitment must ensure you are partners in achieving your customers' goals. Customer focus is better defined as facilitating customers' success. As you will define in sub clauses 8.2.1, 9.1.2, and 9.1.3, open communication and continual feedback with your customers can be a gold mine of information.

Customer dissatisfaction is the most clear and present risk threat any leader can have. Risk avoidance is a critical component of this BMS. Ask. "How will our customer react if product, service conformity, and on-time delivery results are not or will not be achieved? Have we built appropriate processes into our BMS to avoid these risks?"

5.2 Policy

5.2.1   Establishing the quality policy

Traditional quality policies must be superseded with BMS principles, which must be authored by senior management and become integral to daily operations.

A quality policy must contain a mandate for it to be followed absolutely, and remedies for it not being followed. Ask. "Have we established objectives and a means for reviewing them? Are we committed to continual improvement, and are there resources in place to attain this?"

5.2.2   Communicating the quality policy

The BMS principles must become the core of the organization. Communicating them is a requirement of onboarding and ongoing training. Everyone in the organization must live the quality policy every day.

A policy for quality is ineffective unless it mandates that defects never reach a customer. Is your quality policy complete, available to and understood by all interested parties?

5.3   Organizational roles, responsibilities, and authorities

Operational procedures, roles, and responsibilities must be included in a master plan of harmonized documents with measurable outcomes and consequences. An individual who executes any role must be trained and certified as competent before performing that role.

Roles, responsibilities, and authorities must be clear, concise, and enforced. Be aware that the potential for foreseeable risk can be rampant. Ensuring that responsibilities and authorities for relevant roles are assigned, communicated, and understood is critical to risk avoidance.

Ask: "Are you ensuring customer focus throughout the organization? Do you have a way to report risks as they are found? Do you have a process to actively mitigate them? Do your actions intrinsically avoid foreseeable risks?"

6 PLANNING

6.1 Actions to address risks and opportunities

Develop an irrevocable company goal that NO critical defect shall ever reach a customer.

- Use quality process improvement tools to map all business processes, their interrelationships, and their interdependencies.
- Assess each process on its own and assign KPIs for effectiveness, quality, and reliability.
- Dissect each process until all opportunities for defects or mistakes are identified and understood.
- Objectively improve the processes until they are innately safe, reliable, and continually monitored. Ask: Do these metrics warn of pending problems before they happen?
- Purge the concepts of defect inspection, rework, and punishment. Replace them with individual and team accountability for the outcome of everyone's work. Build in the discipline and controls to get the process right the first time. If a potential error or risk is found, do you have a method to report and ensure its correction?
- Continually elicit and scrutinize customer direct and indirect feedback for potential issues.
- Have a clear road map of your business process—where it has been and where it is going. Share that with everyone, regularly.
- Organizations must be aware of their legal Duty of Care, Standard of Care, and Foreseeable Risk.

See Appendix B for a comprehensive risk avoidance tool.

ISO 9001 requires that organizations plan actions to address risks and opportunities. As stated in Clause 4, when defining the infrastructure of the organization and the BMS, create new stretch goals to take your process to the next level.

6.2 Quality objectives and planning to achieve them

- Quality objectives are only effective if they are integral to the overall strategic plan of the organization. Planning to achieve them must be part of the enterprise wide objectives.
- When planning how to achieve quality objectives, you must establish:
- What is the return on investment?
- How will the vision, mission, and values be implemented and measured?
- How will the customer become a referral for us, not a victim of poor quality?
- How will you achieve 100-percent defect-free products and services?
- Have you identified the resources necessary to achieve your quality objectives? (see 7.1 below)
- Have you assigned direct responsibility for each objective?
- How do you identify foreseeable risk?
- How do you avoid foreseeable risk?
- How do you anticipate how customers can misuse your products?

Since many changes are unforeseen, you must build processes for verifying, validating, and implementing changes. You also need a built-in process for evaluating the effectiveness of any changes. This ties directly back to your ROI.

6.3 Planning of changes

A change-management program must be effective for the company, culture, and ultimately, your customers.

While implementing process changes, use the criteria to assess each process. Identify opportunities to add value and identify foreseeable risk. Clearly identify and understand the purpose for a change and any potential consequences. Make sure that potential changes have not been tried before unsuccessfully.

The same risk questions in 6.2.1 and 6.2.2 need to be incorporated in changes. There should be stricter scrutiny of how the changes affect other processes.

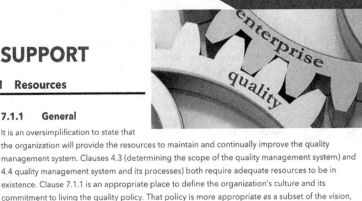

7 SUPPORT

7.1   Resources

7.1.1   General

It is an oversimplification to state that
the organization will provide the resources to maintain and continually improve the quality
management system. Clauses 4.3 (determining the scope of the quality management system) and
4.4 quality management system and its processes) both require adequate resources to be in
existence. Clause 7.1.1 is an appropriate place to define the organization's culture and its
commitment to living the quality policy. That policy is more appropriate as a subset of the vision,
mission, and values of senior management. Without that commitment, understanding, and
associated training, merely providing resources is a plan for mediocrity.

7.1.2   People

Your organization must determine and provide the people necessary for the effective
implementation of the business management system and for the operation and control of
processes. As with 7.1.1, compliance can only be determined by auditing the entire BMS as a
whole, not just through periodic audits of individual functions or departments.

In an effective business management system, each employee must be competent in their often
multiple roles and responsibilities. Competency must be verified on a periodic basis. It should be
tied to reward and disciplinary actions.

People make up most of the risk potential in any organization. When people are unaware of
foreseeable risk and how to avoid it, the organization remains in a state of potential liability. The
organization must implement a risk avoidance program.

7.1.3   Infrastructure

The organization shall determine, provide, and maintain the infrastructure necessary to operate its
processes and achieve conformity of products and services. Unless senior management creates a
culture that encourages excellence you will be destined to mediocrity and risk.

In creating the infrastructure, all tenets of business process excellence and risk avoidance must be
included in the design and implementation. Have you considered buildings, utilities, equipment,
software, transportation, environmental stewardship, and other elements as part of your
infrastructure? Review each of these for risk and take steps to avoid any foreseeable risks. As
stated in 7.1.2, you should include these in your audit of the entire BMS.

7.1.4   Environment for the operation of processes

As in 7.1.3, the organization shall determine, provide, and maintain the environment necessary to
operate its processes and achieve conformity of products and services. Since process

environmental factors are soft skills, senior management must define the environments that are conducive to their organizational needs and provide a productive and cohesive workforce.

An often-overlooked clause of the Standard, 0.4, states: "This International Standard does not include requirements specific to other management systems, such as those for environmental management, occupational health and safety management, or financial management." Do not confuse 7.1.4 with these stated exclusions; include them in the overarching audit of your business management system and look for associated risks. This can directly link to your company's success.

General

This requirement establishes a viable metrology program commensurate with the products or services being manufactured. As with many other clauses, the difference between mediocrity and excellence is that for the latter, the goal must be that no critical defects ever reach a customer.

When sampling programs are employed, they should be evaluated for how successfully they avoid risk. The records of device calibration and the data collected during measurement and test must be scrupulously maintained. They could become evidence in a lawsuit.

Measurement Traceability

Traceability is a fundamental tenet of foreseeable risk. The organization shall provide coherent traceability of any component, subassembly, assembly, or components of service necessary for the organization's goals or as prescribed by outside requirements.

Questions to consider as you audit your BMS and look for foreseeable risks:

- If a product or service fails in the field, do you have a way to verify that it passed your internal measurements or tests prior to delivery?
- If you find a Test, Measurement and Diagnostic Equipment (TMDE) tool out of calibration, can you determine what it was used on prior to finding it out of calibration?
- Do you have a viable recall process?

7.1.5   Organizational Knowledge

The organization shall determine the knowledge necessary to operate its processes and achieve conformity of products and services. The tenets of 7.1.6 must be the call to action for senior management to define their company's culture and assess how each of these requirements can contribute to business-process excellence. Your competitors are always looking for an advantage. So, in addition to reviewing your successes, you should also look at past failures and learn from them. Remember, a company (or process) that doesn't learn from both successes and occasional failures is one that doesn't advance.

7.2 Competence

Determining the competence of every individual within the BMS is critical to attaining business process excellence and risk avoidance. This is best accomplished by establishing roles and responsibilities, rather than job descriptions.

Evaluating competence must be an ongoing process. Your people thrive in an environment where they know they can advance. In addition to initial onboard training of your people, you should develop a robust continuing training program that focuses on company and business-process excellence. Empower your people to identify improvement processes and associated risks.

While necessary from an HR and potentially legal or disciplinary perspective, many job descriptions include the phrase "and all other duties as assigned." It is impossible to determine competence with such a vague requirement. Roles and Responsibilities must include actions to be taken and an expected outcome. The outcomes must be quantifiable.

Determining competence takes more than annual performance reviews. It must be an ongoing activity, monitored continually. It cannot be a façade of compliance.

A tool to determine competencies is including that function during internal audits. The auditee will demonstrate competence in the answers given during interviews.

Individuals can hold multiple competencies that are continually monitored.

7.3 Awareness

The organization shall ensure that people doing work under the organization's control are aware of all requirements of the BMS. Awareness must be a continual activity, not just formal meetings, bulletin boards or internet sites. Scrolling reminders on monitors can raise awareness, but it must be complemented by human interaction.

Ask: "Are my people aware of their contribution to effectiveness to the BMS? How can they improve the quality and safety of your product/service? Do you stress the importance of the defined business ethics that tie directly to your standards of care and accountability?"

7.4 Communication

The organization shall determine the internal and external communications relevant to the business management system. Communication must be an ongoing activity through meetings, OJT, and open-ended dialogue. Ensure that your people are welcomed to bring any concerns they may have to senior leadership. This is often one of the best ways to discover improvement opportunities.

The focus of communication is to reinforce awareness and the vision, mission, and values of your company. You should also continually communicate the need for safety, for your people and for the customer.

Validate communication to avoid miscommunication.

7.4.1 Documented Information

The first requirement is to ensure documentation required by the standard is created. This is a compliance matter that segues into the second requirement.

The second is to create documentation that gives clear and concise direction for the processes and the competence level of the user. Well-documented processes and specifications are key to organizational excellence.

The extent of documented information is directly related to the complexity of the organization, its products, and services. The culture of the organization must be reflected in the documentation.

Absence of needed documented information is a major component of organizational risk.

7.4.2 Creating and updating

Creating and updating documentation requires a formal system of creation, review, and approval. Databases must be secure, with only the current revision of documents available to users.

Notification of new and revised documents must be given to the appropriate stakeholders, and there must be an acknowledgement that these documents have been received, understood, and communicated.

Documents must have a dated timestamp and a policy describing how long a document can be used and how to dispose of it after printing.

There must be a policy of how many revisions are permitted before a document should be rewritten. While it's generally understood that redlines are necessary, a policy is also needed that describes how a redlined document (and the number of redlines on it) can be used on the shop floor before it is submitted for revision.

Control of Documented Information

Documented information required by the business process management system and by the Standard shall be controlled to ensure integrity and the distribution process. Servers and other computerized systems should be automated so only the latest revisions are available with historical revisions accessible only to those who have the need to know. Another tool for excellence is to hyperlink the documentation to subordinate documentation to prevent data-entry errors.

Control must include retention and disposition policies with adequate protection from loss of confidentiality, improper use, or loss of integrity. Remember that your company's reputation is on the line. Your documentation is your intellectual property, and lack of proper control can be risky and costly.

There are many document control programs available to ensure security.

End of free abridged version.

Appendix G: Quality as a Profit Center

QUALITY AS A PROFIT CENTER—A CASE STUDY

DELL ARB

Initiatives, which some would justifiably call experiments, that attempt to treat quality management in a strategic business way are not new. Since the creation of the ISO 9000 standards in the 1990s, there have been numerous attempts to envision quality management as a strategic entity of the business, not simply a compliance exercise. This case study is one of those attempts; a successful one.

Dell implemented Quality as a Profit Center (QPC) to its offerings to highlight the need to use the Standard not just as a conformance standard but as an opportunity to define a return on investment (ROI) for each Clause of the Standard. The first full-scale implementation was at a division of Dell Computer. The case study will provide a more concise example of QPC.

The Asset Recovery Business of Dell (Dell ARB) became a business unit in February 1999. Dell's total satisfaction return policy spawned an influx of computers and laptops that were returned for reasons ranging from buyer's remorse to the customer just not liking them. This clever marketing strategy created a massive customer following that was revered in its heyday.

The first facility was a decommissioned department store in Austin, Texas. When we first arrived to assess the ISO 9001:2000 transition project, we observed a daily lineup of customers outside Dell's retail outlet waiting to discover the array of computers in their used inventory that had been refurbished the day before. These were mostly bargain hunters looking for almost-new computers at discounted prices. The refurbished inventory was typically depleted within a few hours of opening the door.

The first project was to create an online portal to aid in managing the business, involving everything from inventory to work instructions. We called it BMIS, or Business Management Interactive System. As BMIS was launched BMIS, a disturbing trend began evolving. Customers were returning new computers, but they were missing hard drives and memory chips. As we began developing the implementation plan, the first order of business was to set up a test station for receiving and denying refunds to those sending back scavenged computers. Process improvement investigations often uncover foreseeable risk.

Instead of creating a quality policy, we incorporated a directive from senior management to guide our implementation planning. This was an early example of creating an enterprise-wide business management system.

LESSONS LEARNED

SENIOR MANAGEMENT MANDATE: TRANSITION

The contract was won to do this transition by continually reinforcing the methodology of QPC (Quality as a Profit Center). Dell management immediately saw that there was no need for an overhead QMS, but instead an enterprise-wide BMS. This was illustrated by senior management's mandate:

- ARB will be successful only by focusing on the management system as a whole.
- ISO 9000–1994 focuses on operations related to production and delivery of goods and services.
- ISO 9000–2000 scope will require an enterprise-wide focus.
- The ARB Business Management System will be designed to address current and future management system requirements by incorporating a continuous process improvement methodology across the enterprise.

SENIOR MANAGEMENT MANDATE: BMIS DESIGN CRITERIA

A uniform web presence was created for each screen of the BMIS. All references to ISO were to be eliminated. Senior managers were each responsible for the content of the BMIS.

- All processes, procedures, and work instructions will follow the same model.
- All ARB documentation will be developed for dissemination via the intranet.
- Documents will be in native html format. Enabler applications will not be called to open documents.
- Revision and version control will be accomplished by use of intranet publishing software and Dell control mechanisms.
- All references to "ISO" are being eliminated from our business documents
- "If it isn't right for our business, we won't do it."
- ARB senior managers are responsible and accountable for the content of the BMS.

SENIOR MANAGEMENT MANDATE: SIMPLICITY IS KEY

A straightforward and simple process flow was created for the BMIS Process Flow (see Figure G1.1). Clicking on any icon would take you to the appropriate screen.

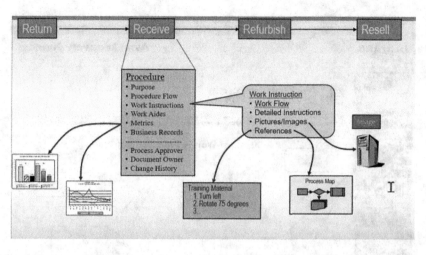

FIGURE G1.1

SENIOR MANAGEMENT MANDATE: RAPID IMPLEMENTATION AND VALIDATION

The BMS must be ready for third-party auditor review in nine months. It was believed that an aggressive schedule was critical for maintaining momentum.

- March 2000: Partnership between ARB and Productivity Resources, and to design and implement the ARB.
- Business Management System Business Management Interactive System (BMIS).

Phase 1 implemented in May 2000

- Dedicated Intranet presence.
- 43 Processes and Procedures.
- BMIS Phase 2 developed and deployed to 85% of ARB by December 2000 NSAI review.
- Web-enabled and linked processes, procedures, work instructions, aides, and reference documents.

SENIOR MANAGEMENT MANDATE: EASE OF USE

This is a typical BMIS web page (see Figure G1.2). Access to procedures, work instructions, and configuration tools must all be less than three clicks away from the current screen. The quality policy and mission statement were highly visible.

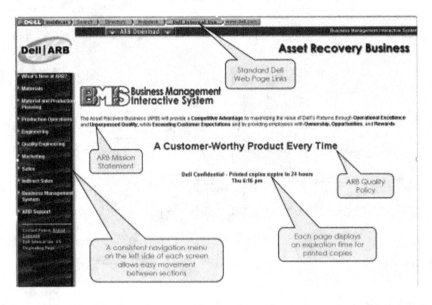

FIGURE G1.2

Senior Management Mandate: Usable "at a glance"

This is a typical operational procedure. A "box and bullet" design for procedures was created as shown below in Figure G1.3. It included process steps and required outcomes. A similar design was used for roles and responsibilities. Metrics for each were one click away.

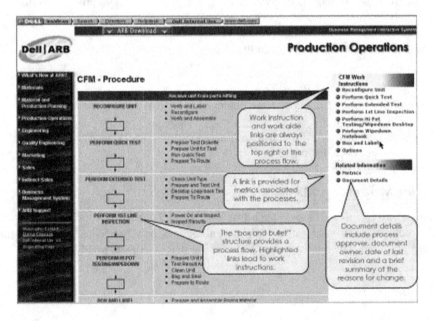

FIGURE G1.3

RESULTS

REGISTRAR'S EVALUATION

For anyone concerned about compliance with ISO 9001 when BMS removed all references to the Standard and did not relate each section to clause numbers, the following quote from their registrar should dispel any fears.

REPORT ON THE SURVEILLANCE AUDIT OF THE QMS OF ARB

NSAI—15 DECEMBER 2000

"The web-based document system recently produced by ARB is approaching best in class in the industry . . . . It is considered that this process . . . should be captured as best Dell Practice . . . The fact . . . that it includes Sales, Finance, Engineering (etc.) is considered a major plus."

Even more compelling are the data from the first year's business results. These are the points you can use to convince your senior managers or consulting clients that an enterprise-wide BMS is far superior to an overhead QMS in achieving the strategic goals of the organization.

AFTER ONE YEAR

One year after the ISO 9001:2000 transition, the Division VP of Dell reported the following:

- Led business from a significant operating loss toward breakeven/profitability. Developed the business to sustain 40–50% year-over-year growth and integrated and aligned the business with the larger Dell segments as it approached $1 billion in sales and asset management.
- Realigned the business to accomplish a projected net savings of more than $100MM.
- Established a Web-enabled Quality Business Management System based on 0W-ISO 9001:2000 to map business processes, communicate business initiatives across the company, cross-train the employee base, and report daily progress against business metrics.
- Increased productivity from $433,000 to over $581,000 of asset revenue per employee, with headcount managed down to 690 employees from a planned 920-employee headcount.
- Launched E-Commerce sales channel with penetration rates of 52% in the consumer segment and 31% in the business segment.
- Reduced Marketing and Sales costs to $23/box, from $101 to $78/box.
- Increased unit sales 145% year over year in the consumer segment and 106% year over year in the commercial segment.
- Increased segment sales 611%, year over year.
- Consolidated five manufacturing facilities into two, with a plan to merge into one facility by 2Q02.

- Established a $780MM manufacturing capability, with factory velocity increasing from 1,000 units/day to 1500 units/day, with a 4Q01 record of 1715 units/day.
- Reduced OBA/OBE defect rate from 28% to 7.3%, with MW&D reduced from 4.6% to 2.5%.

Dell ARB is now www.dellrefurbished.com, where you can select the refurbished products that best suit your needs. This business model has stood the test of time for more than 20 years.

Dell co-authored a book about this methodology, which became one of the founding studies in evolving a QMS into a BMS. As more and more organizations are questioning the value of their ISO 9001 certification, they are also asking a critical question: We know we need a framework for quality management, so what do we replace it with? Perhaps the answer is that business operations have evolved to the point that we now need not a quality management system with all its inherent limitations, but a business management system and quality will be taken care of—because it need no longer be seen as a cost center, but a profit center.

Appendix H: The Risk Avoidance Handbook

Effective Risk Avoidance

A BREAKTHROUGH APPROACH TO RISK MANAGEMENT AND RISK-BASED THINKING

INTRODUCTION

The Oxford Dictionary defines risk as a situation involving exposure to danger. According to the Risk Management Institute, risk management involves understanding, analyzing, and addressing risk to make sure organizations achieve their objectives.

ISO/TC 176/SC2/N1284 guidance document for ISO9001:2015 suggests that risk-based thinking is intended to establish a systematic approach to considering risk, rather than treating "prevention" as a separate component of a quality management system. It goes on to state that risk-based thinking is something we all do automatically in everyday life.

ISO 31000:2018 states that the purpose of risk management is the creation and protection of value. In the eight principles outlined in the standard, one tenet includes risk management, which anticipates, detects, acknowledges, and responds to changes.

The quality profession is embracing risk management as an evolution of preventive action. Unfortunately, the same term, risk management, is an entire industry that primarily focuses on statutory and regulatory compliance.

The gamut of definitions of "risk" runs from something we avoid automatically in our daily lives to avoiding danger in business, protecting value, adhering to regulations, and avoiding a lawsuit. How do enlightened business leaders decide the definition of risk for their organizations? How do quality professionals provide guidance for risk abatement in those organizations? How do organizations immunize themselves from the potential of civil litigation?

In legal parlance, "risk" is the basis for products liability and organizational negligence tort. My work as an expert witness uses quality standards auditing techniques to prove an acceptable or negligent standard of care by a defendant company.

The definition of risk-based thinking is something we all do automatically in everyday life, is an oxymoron. As quality professionals, we would be out of business if that were true.

RISK AVOIDANCE

My experience as a quality control engineer and a testifying expert witness has caused me to conclude that the creation of defects that lead to exposure to danger is avoidable. Risk is avoidable because it can be foreseeable.

Before I witnessed the outcome of process variability, such as catastrophic fires, injury, loss of property, and death, I was not aware that our traditional approach to quality management was fundamentally flawed. Traditional preventive action is often triggered by a problem being foreseen in one or more processes. Those who employ tools such as FMEA's can uncover potentially detrimental process inconsistencies in their investigations. Those who rely on PAR's are not proactive enough to practice risk avoidance. Rarely is preventive action a result of an epiphany or flash of genius. Nor is it a result of cognitive risk awareness.

Even more uncommon is a formal program of foreseeable risk implemented as an immutable and proactive cultural mandate.

FORESEEABLE RISK

Businesses are living organisms. Each company exhibits a unique personality and has attributes distinctive to its own heritage and lifestyle. Organizations possess a measurable state of wellness that is in constant flux depending on internal fitness and the relentless assault of outside pathogens.

Business health can span the range from the peak performance of a world-class contender to being terminally ill. Overt appearance is seldom an indicator of a state of well-being at any given moment. Symptoms of the disease may be known and treated and are manifest but are being ignored or undetected, growing and consuming the organism.

Foreseeable Risk is a scientific diagnostic methodology that examines the health of a business utilizing proven tools of process analysis, performance to standards, business metrics, and management system effectiveness. The first step of triage is a detailed workup of the overall health of the business backed by clinical evaluations of each constituent element of the company.

Used proactively, Foreseeable Risk is a roadmap for organizations to achieve peak health and wellness. Used as a forensic tool, it provides documented evidence of standard of care and foreseeable risk measured against quantifiable standards of performance.

For enlightened business leaders, Foreseeable Risk is a strategy for achieving peak performance while immunizing the company from products liability and organizational negligence. By virtually eliminating product defects and service errors, organizations can achieve unparalleled pinnacles of customer service.

ADOPTING THE RISK AVOIDANCE PARADIGM

For organizations to be successful in risk avoidance, there must be a top-down change in basic assumptions in the vision, mission, and value system of the company. Whether your company culture has morphed organically over time or is the result of enlightened strategic planning, the concept of risk avoidance is never part of the company mission.

The fault is not one of pragmatic process development or selecting one quality management approach over another; it is embedded in the fundamental paradigms of organizational leadership. Risk avoidance is not taught in B school or in management seminars. It is not part of business or quality management training curricula.

It is a new paradigm that states that defects are avoidable, not inevitable. If your value system is based on the inevitability of some level of nonconformities being acceptable, you will always have nonconformities.

Unfortunately, since ISO 9001:2015 added risk-based thinking to our lexicon, quality professionals have attempted to create compliance parameters for risk-based thinking rather than taking a 10,000-foot level assessment of its meaning and implications. The following steps are not found in your father's quality management system.

1. Revisit your Vision, Mission, and Values
   a. Senior management must become fully aware and enlightened on the tenets of risk avoidance and foreseeable risk.
   b. If you are to be successful, the vision, mission, and values must be revised to include a mandate for risk avoidance.
   c. The new mandates must be communicated at all levels and incorporated as a culture, not a requirement.
2. Reassess your quality management system
   a. While ISO 9001:2015-based QMS addresses business success, implementation is typically founded in the traditional steps of continually improving processes to achieve more favorable outcomes and metrics.
   b. Determine where your organization falls in Table 16.1 and determine for which level you are initially striving.
   c. The QMS reassessment process must be one of changing the focus to quality as a profit center, stupid proofing processes, and products, metrics driving key business indicators, avoiding defects reaching the customer, and achieving perfect customer report cards.
   d. Process control must include the inculcation of the vision, mission, and values.
   e. Each tenet of your QMS must be restated from risk management to risk avoidance guidance.
   f. There are proven tools that have been developed for risk avoidance.
3. Implement the Risk Avoidance business model
   a. Conduct presentations and workshops to teach Foreseeable Risk.
   b. Assess your compensation program. Consider a risk and reward program that has incentives for defect-free performance and penalties for creating nonconformities.
   c. Recast your version of PDCA to replace the "check" function with an "avoidance" function.
   d. Assess the control limits of your metrics to strive for zero nonconformities, which is not an acceptable number.
   e. Utilize every defect report, customer input, and corrective action as gifts for your new goal of delivering defect-free products and services.

THE TOOLS OF FORESEEABLE RISK

Assessing foreseeable risk is the next level in quality management. It is not reinventing your business processes or adding more overhead to ensure conformance.

Before delving into the tools, take a moment to study. It is included to help with a change in thinking from the requirements of ISO 9001 to a business-level view of risk avoidance.

1. Assess and revise the steps of each process in your organization. Include two mandatory questions for the process operator to answer before moving the product or service to the next process step.
   a. Am I certain that my work is defect-free before passing it to the next process step?
   b. Am I solely accountable to the next process operator for my work product?
2. Add a step to audit and management review meetings. Ask the questions.
   a. How can the owner of the next process use my work output badly or inappropriately?
   b. How can the work output of each process affect a customer?
3. Add a step to your FMEA processes or impanel a risk avoidance committee. Include individuals who have no formal involvement in the processes but are dedicated to the success of the company. Add these steps:
   a. During DFEMA's and PFEMA's add a round of questioning about how a customer could use the product or service badly, stupidly, or for the wrong purpose.
   b. Review customer complaints, warranty reports, and repairs with a deliberate focus on what we could have done to avoid each issue. Customer stupidity will not be a relevant factor in a lawsuit if you have not adequately anticipated misuse.
4. Continually model the culture that a defect will never leave my workstation; therefore, a defective product or service will never reach a customer.

SUMMARY

- Develop an irrevocable company goal that NO critical defect shall ever reach a customer.
- Use quality process improvement tools to map all business processes, their interrelationships, and their interdependencies.
- Assess each process on its own and assign metrics for effectiveness, quality, and reliability.
- Dissect each process until all opportunities for defects or mistakes are identified and understood.
- Objectively improve the processes until they are innately safe and reliable, and continually monitor metrics to warn of pending problems before they happen.
- Remove from your organization the concepts of defect inspection, rework, and punishment.

- Replace them with individual and team accountability for the outcome of everyone's work.
- Continually elicit and scrutinize customer feedback for potential issues. Correct them immediately and proactively as a "gift" of valuable information.
- Have a clear roadmap of your business, where it has been, and where it is planned to go. Share that with everyone regularly.
- Organizations must be aware of their legal Duty of Care, Standard of Care, and Foreseeable Risk.

CALL TO ACTION

The only human who embraces change is a wet baby. Proposing that organizations adopt fundamental culture shifts is right up there in pain intensity with a root canal.

For most of my career, I have worked with organizations that see quality management as an overhead expense that requires a police force to be effective. For business leaders, it is too often mandated by customer requirements or as a fix to product and customer issues. For quality professionals, it is a culture of tools and certifications that exist as the foundation for their very existence.

Business leaders often see new concepts such as those proposed herein as more bureaucracy bubbling up from the quality department. My fellow quality professionals: If we, as a profession, do not refocus our energies and work on creating customer satisfaction, increasing return on investment, and holding everyone accountable for their work output, our litany of quality tools will be forever reactive instead of proactive.

One way for us to get the undivided attention of senior management is to sort through the morass of definitions of risk at the beginning of this article and synthesize a culture of risk avoidance that is unique to your organization. It must replace preventive action, risk assessment, risk management and incorporate standards, statutes, regulations, and industry norms into the mainstream workflow.

Seek out every opportunity to remove traditional overhead and tollgate functions and incorporate them as process steps that have a positive ROI. In doing so, you will increase your value to the organization as an expert in avoiding expensive customer defect issues. You may also create a new function as an internal expert witness in products liability and organizational excellence.

ASSESSMENT QUESTION RISK FACTORS

Have you communicated the core values and vision of the founders to all members of the organization and then enforced them?

As organizations grow, the core values of the principals become diluted and often forgotten. Expeditious business decisions may introduce unexpected and undetected deterioration in product and service quality.

Are all business processes clearly defined and expected outcomes measured? The seeds of products liability enter your product or service during their construction. Without continual monitoring of each process step, the final product may include concealed opportunities for failure.

Are your customer service people always busy? Customer calls, returns, and warranty claims contain a wealth of information about pending disasters. Each event must be scrutinized, and common issues must be investigated thoroughly for their root causes with the goal of driving warranty claims to extinction.

Have the older customers of your organization stopped buying from you?
Customer loyalty is fleeting, and deteriorating quality or service may motivate customers to abandon you without notice or cause. Investigate why ex-customers have stopped calling.

Appendix I: Conducting Competence Assessments

Quality Masters' Training Program

DEFINING COMPETENCY

DEFINITIONS OF COMPETENCY

- Possession of required skill, knowledge, qualification, or capacity.
- Having suitable or sufficient skill, knowledge, experience, etc., for some purpose; demonstrably qualified.
- Meeting certain minimum requirements.
- Adequate skills, but not necessarily exceptional.
- Possession of required skill, knowledge, qualification, or capacity.
- By training and experience, an individual possesses a manual or soft skill that can be repeated continually without innate errors.
- That person has working knowledge of every aspect of the skill level being evaluated.
- There are a set of standards established by which competency in that skill can be benchmarked and evaluated.
- Industry Standards.
- Company Standards.
- Regulatory Requirements.
- The individual has the mental acuity to anticipate the variations that may occur during the performance of that skill and to knows when to ask for help.
- Having suitable or sufficient skill, knowledge, experience, etc., for some purpose; properly qualified.
- Skills to be evaluated must have clearly defined bounds of acceptability.
- Skill levels must also be clearly defined.
- The skill must have a clear and concise purpose and parameters of how the product or service step interacts with its upstream and downstream processes.
- "Qualified" must contain unambiguous criteria based on well-conceived and documented roles and responsibilities and work instructions.

NOTE: Speaking, reading, writing, and comprehending basic English language skills MUST be a core competency.

- Being competent is defined as:
- Meeting a set of minimum requirements for a particular skill level,

- Meeting all minimum requirements,
- Demonstrating competency to another individual who is certified at a higher level of that skill,
- Competency is adequacy, not being exceptional,
- A task can be completed with minimum acceptable results every time.

PREREQUISITES FOR DETERMINING COMPETENCY

- Classifying the skill.
- Defining the required level of formal training.
- Defining the required level of experience.
- Creating the tests to determine that the prerequisites have been met.
- Defining the go-no-go parameters for determining whether the prerequisites are genuine.
- Keeping the evaluation process objective and uncompromised.
- Basic: Tasks are rote, repetitive, and do not require extensive knowledge or training.
- Typically, manual skills like janitorial or file clerk.
- Process Operator: An individual who has certain skills that enable them to competently operate a predefined set of tasks that have unambiguous outcomes.
- Typically, someone repeatedly performs tasks that require manual dexterity and a basic understanding of the product or service.
- Also required for a probationary employee.
- Needs Work Instructions to perform detailed tasks.
- Level 1 Competency:
- An individual who has demonstrated a basic level of comprehension of what the tasks are and why they are important to creating zero defective products.
- Demonstrated knowledge and understanding of the mechanics of performing the process and why each step is important to success.
- Demonstrated dexterity whether it is manual or a soft skill.
- Able to show compliance with all required competencies in random audits.
- Completion within a defined time limit where work output consistently meets all task requirements.
- The demonstrated ability to be accountable for their own work to themselves and their team members.
- Level 2 Competency:
- An individual who has demonstrated a detailed working level of comprehension of what the tasks are and how they are important to creating no defective outcomes.
- The knowledge to assess and improve the steps of performing the process and why each step is important to continual improvement.
- Demonstrated operational expertise whether it is manual or a soft skill.
- Being able to assess Level 1 competencies in performance audits.
- Work output is consistently above minimum requirements.
- Demonstrated leadership and mentorship skills.

- Level 3 Competency:
- An individual who has demonstrated a performance level that exemplifies craftsmanship.
- An overall deep working knowledge of the processes and their interactions.
- Creating a benchmark of process performance excellence.
- A skilled observer and auditor in all aspects of the processes.
- Work output sets the standard that all others should work toward.
- Demonstrated ability to teach, mentor, and document process procedures and create more effective processes.
- Demonstrated capacity and passion to be certified in more advanced skills.
  Sample Roles Requiring Competency Criteria
  Sample Skills Requiring Competency Measures
  Creating a Competency Matrix (Sample)
  Creating Competency Verification Checklist (Sample)
  Establish Evaluation Criteria
- Possession of required skill, knowledge, qualification, or capacity.
- Having suitable or sufficient skill, knowledge, experience, etc., for a defined purpose; properly qualified.
- Meeting certain minimum requirements.
- Adequate but not exceptional.
- Continual Evaluation and Corrective Actions.
- Mandatory ongoing process administered by senior staff
- Weekly requirements of their job.
- Mining data from NCRs, CARs, and PARS to validate competencies and needed corrections.
- Evaluating competency during internal audits.
- Tiger Teams to solve individual issues.
- Under QMS review for continual improvement.

Index

213

Printed in the United States
by Baker & Taylor Publisher Services